AF182411

Toward Consilience

Toward Consilience

The Bioneurological Basis of Behavior, Thought, Experience, and Language

Gerald A. Cory, Jr.
Center for Behavioral Ecology
San Jose, California

Springer Science+Business Media, LLC

Library of Congress Cataloging-in-Publication Data

Cory, Gerald A., Jr.
Toward Consilience: The Bioneurological Basis of Behavior, Thought, Experience, and Language/Gerald A. Cory, Jr.
p. cm.
Includes bibliographical references and index.
ISBN 978-1-4613-6918-9 ISBN 978-1-4615-4271-1 (eBook)
DOI 10.1007/978-1-4615-4271-1
1. 2.

ISBN 978-1-4613-6918-9

© 2000 Springer Science+Business Media New York
Originally published by Kluwer Academic/Plenum Publishers in 2000
Softcover reprint of the hardcover 1st edition 2000
http://www.wkap.nl/

10 9 8 7 6 5 4 3 2 1

A C.I.P. record for this book is available from the Library of Congress

All rights reserved
No part of this book may be reproduced, stored in a retrieval system, or transmitted in any form or by any means, electronic, mechanical, photocopying, microfilming, recording, or otherwise, without written permission from the Publisher.

To my three sons
Daniel, Donald, and Gerald III

PREFACE

The present work is the second in a series constituting an extension of my doctoral thesis done at Stanford in the early 1970s. Like the earlier work, *The Reciprocal Modular Brain in Economics and Politics, Shaping the Rational and Moral Basis of Organization, Exchange, and Choice* (Plenum Publishing, 1999), it may also be considered to respond to the call for consilience by Edward O. Wilson. I agree with Wilson that there is a pressing need in the sciences today for the unification of the social with the natural sciences. I consider the present work to proceed from the perspective of behavioral ecology, specifically a subfield which I choose to call interpersonal behavioral ecology

Ecology, as a general field, has emerged in the last quarter of the 20th century as a major theme of concern as we have become increasingly aware that we must preserve the planet whose limited resources we share with all other earthly creatures. Interpersonal behavioral ecology, however, focuses not on the physical environment, but upon our social environment. It concerns our interpersonal behavioral interactions at all levels, from simple dyadic one-to-one personal interactions to our larger, even global, social, economic, and political interactions.

Interpersonal behavioral ecology, as I see it, then, is concerned with our behavior toward each other, from the most obvious behaviors of war among nations, to excessive competition, exploitation, crime, abuse, and even to the ways in which we interact with each other as individuals in the family, in our social lives, in the workplace, and in the marketplace. It is about more, however, than just damage control, adjustment, and repair to the structure and behavior of our interpersonal lives. It seeks to go further – to understand and apply the dynamics of interpersonal behavior with a view to improving the larger social, economic, and political systems that shape our lives.

This present book seeks to identify and explore the basic algorithms of our evolved brain structure that underlie our thought, behavior, language, and subjective experience. In the quest to unify the social with the natural sciences, we must inevitably turn to evolutionary neuroscience as the bridging discipline. There is no where else to go. Although our brain evolved under constraints of the laws of physics and chemistry, the evolutionary process itself involved chaotic and random factors, as well as natural selection processes. The algorithms of our brain, then, which are the foundation of our social sciences, can never have the immutability and predictability of the laws of physics and chemistry. These algorithms, although dynamic, shaping factors of our behavior, are, to a degree, innately variable and experientially modifiable. This fundamental difference between the natural and social sciences precludes a simplistic reduction, but indicates, rather, the establishment of linkages and bridges.

ACKNOWLEDGMENTS

The ideas presented in this book evolved over a period of a half-century during which I accumulated many intellectual debts. My early interest in the brain goes back to my employment at Tidewater Hospital, a private psychiatric institution (1950-1951) in Beaufort County, South Carolina. Under the supervision of psychiatrist, A. K. Fidler, I participated extensively in patient care and studied the greats of psychiatry and brain science. I have pursued my interest in neuroscience unflaggingly since that time. I also wish to acknowledge Robert H. Wienefeld, former chair of history at the University of South Carolina, who, during my undergraduate years in the early 1950s helped me to begin thinking historically and economically.

I wish especially to thank Kurt Steiner of Stanford University for his guidance and encouragement. His support of the ideas and concepts of this work has been of great value to me over the years since we first met at Stanford in 1970. Also I wish to thank Robert North, Nobutaka Ike, Hans Weiler, Peter Corning, Donald Kennedy, Charles Drekmeier, Tilo Schabert, Alexander George, Richard Fagen, Daryl Bem, and Robert Packenham, all of Stanford University, who gave their advice and support in the development of these ideas many years ago. I owe special and more recent thanks to Paul D. MacLean of the National Institute of Mental Health, for his review, helpful comments, and continuing encouragement; and to Elliott White, now emeritus of Temple University, for his review of a related manuscript, his encouragement, and his many valuable comments. I also wish to thank Edward O. Wilson of Harvard University for his acknowledgment of my work and his encouragement.

For their review and endorsement of an earlier, related work, I wish to thank Russell Gardner, Jr., emeritus professor of Psychiatry, University of

Texas; Daniel Levine, Department of Psychology, University of Texas Arlington; Kent Bailey, Department of Psychology, Virginia Commonwealth University; Alice Scheuer, Department of Psychology, University of Hawaii (and the WHO Field Psychiatric Center); Brian Barry, Department of Psychology, Rochester Institute of Technology; Andre Fiedeldey, Department of Psychology, University of Pretoria; Arthur Sementelli, Department of Political Science, Stephen Austin State University, Nacogdoches, Texas; and Trudi Miller, Department of Political Science, University of Minnesota. Paris Arnopoulos, of Concordia University, who shares an interest in algorithms bridging the natural and social sciences, also generously reviewed an earlier related work. I wish also to thank Richard Mapplebeckpalmer, Rector of Grace North Church, Berkeley, California for the opportunity to present some of my thoughts to the challenging group at the Grace Institute for Religious Learning and for his encouraging review of an earlier manuscript form of this book.

Vic and Adrienne Hochee, John Lightburn, David McKenna, and Peter Lynch were my valued colleagues over many years and provided their criticisms and support with generosity and thoughtfulness. My thanks also go to my many seminar colleagues and students, who were an inspiration and a challenge over the years.

Michael Hennelly, senior editor at Plenum, was my patient guide and supporter through the production of this book.

As appropriate, I remain responsible for any errors or misinterpretations that may appear in the present work.

Permissions to quote beyond fair use have been granted as follows: Oxford University Press for permission to reproduce fig. 3-2 (p. 32) from Werner R. Lowenstein, *The Touchstone of Life*, Oxford University Press (1999) (copyright 1998). The American Journal of Physics and author Henry Pierce Stapp for permission to quote from "The Copenhagen Interpretation," *American Journal of Physics*, 1972, 40 (8), 1109. Princeton University Press for permission to quote from Richard Rorty, *Philosophy and the Mirror of Nature*. Princeton University Press (1979), 241, 357.

CONTENTS

Toward Consilience

Chapter 1

INTRODUCTION

This book is an effort at bridging disciplines. It responds to the recent call by sociobiologist Edward O. Wilson for consilience (1998), a concerted effort toward unifying the natural and social sciences. This is the second of two books in this bridging effort. The first, *The Reciprocal Modular Brain in Economics and Politics: Shaping the Rational and Moral Basis of Organization, Exchange and Choice* (Cory 1999) dealt primarily with the linkages between economics and political science and, to a lesser extent, with linkages to sociology and anthropology. This book focuses on linkages with thought, behavior, subjective experience, and language. In doing so, it examines aspects of psychodynamics, philosophy, mysticism, ethics, and linguistic theory.

The essential discipline to span the chasm separating the natural sciences from the social disciplines is evolutionary neuroscience, specifically the study of the human brain. Although all things begin with the laws of physics, the evolutionary process that produced the brain interjects random and chaotic elements that deny prediction on a simplistic reductive basis. The human brain, then, is a product of a long period of evolution, the end product of which could not have been predicted from knowledge of the laws of physics and evolution itself. In full recognition of the complexity of the evolved brain, Ramon y Cajal, famed neuroscientist of the late 19th and early 20th centuries, reportedly said that we can never understand the universe until we understand the human brain which created that universe.

Accepting neuroscience as the bridge between the natural and social sciences and disciplines considered does not involve a reductionist program that shrinks all such disciplines down to fit the terms of physics and biology. Scholars in these various disciplines have rightly resisted the suggestion of such simplistic reduction. The bridge of neuroscience establishes the anchors and the linkages for unification. The social disciplines build upon these anchors and linkages, extending them and introducing entirely new and necessary concerns from their unique perspectives and levels of analysis.

Chapter 2 lays the basis for the analysis that follows. All significant experience of humankind is processed through the brain. This is true whatever the source of the experience...internal or external to the body. The brain processes this experience through structured, genetically prewired procedures called algorithms. This chapter examines the origin and nature of algorithms, tracing their usage from mathematical and computer applications to the more mundane activities of everyday life. It sets the challenge of identifying the major generalized algorithms of the brain which process and shape the thought, behavior, subjective experience, and language of our daily lives.

Chapters 3 and 4 identify and define the primary bioneurological or brain algorithm from the perspectives of physics and biology. In physics, the primary algorithm is seen as emerging from the temporary holding or structuring action displayed by living organisms against the second law of thermodynamics, which law moves toward increasing disorder, loss of availability of useful energy, or chaos. In ***formal*** terms of information theory, the algorithm is seen as the self-structuring, self-replicating cybernetic loop of protein-nucleic acid synthesis. At a higher level of generalization, and without specification of informational content and cybernetic controls, the algorithm is expressed ***informally*** as a continuing process of ***order → chaos → order***, as the organism continuously imposes its own informational structure on environmental materials and energy as long as it lives.

From the perspective of behavioral biology, the primary algorithm is proposed as the incorporative aspect of the protein-nucleic-acid synthesizing loop and its related metabolic activities. Behaviorally, organisms maintain their essential structure by reaching or ***extending*** themselves into the environment and ***incorporating*** the necessary energy and materials. This organic process of ***incorporation → extension → incorporation*** is the biological behavioral analogue of the ordering pattern from physics. Both patterns describe the behavioral algorithm necessary to the protein-nucleic acid synthesizing loop as it interacts with the environment to build and maintain its living structures and processes.

The primary algorithm establishes the subject/object behavioral relationship with the environment. As the cybernetic algorithm of biosynthesis, its looping, synthetic features are later replicated in our cognitive neural structure, where it produces a ubiquitous, pervasive shaping effect across the spectrum of human thought, behavior, subjective experience, and even language.

Chapters 5 and 6 develop the second brain algorithm...the reciprocal algorithm of our evolved brain structure. These chapters set out the new modular model of behavior called the conflict systems neurobehavioral model (Cory 1999). This model, of interconnected and distributed modules, draws upon some of psychologist Abraham Maslow's insights while building primarily upon the work of neuroscientist Paul MacLean. Two master inclusive modules are proposed. The first consists of self-preservational programming and is based primarily, although not exclusively, in the tissues of earlier brain structures which were common to early ancestral amniotes, reptiles and mammals. The second consists of affectional programming which is based primarily, although not exclusively, in the tissues of the brain which became highly developed with mammals and produced the peculiarly mammalian affectional behaviors of maternal nursing and long-term infant-parent-family bonding.

These two master modules are driven by our cellular and body processes of metabolism mediated by hormones, neurotransmitters, and neural architecture. They act dynamically to shape our behavior in the environment and set us up for a life of internal as well as external conflict because of their often conflicting priorities. Behavioral tension occurs when the two are in conflict or frustrated in behavioral expression. The often conflicting urges of these two master modules are input by way of bi-directional and multi-lateral neural pathways to the more recently evolved neocortex, which is capable of language and thought. These urgings are represented in the neocortex (a master language, thought, and executive module), and are expressed at a high level of cognitive generalization as ego (self-preservation) and empathy (affection). The executive function of the neocortex (especially the frontal cortex) has the capacity and the responsibility for making our rational and moral choices from among these often conflicting behavioral priorities. The documentation of these three master modules draws upon a wide spectrum of neuroscience, ethology, and psychology (cognitive, motivational, and attachment). Although, like the primary algorithm, the second is dealt with at a high level of generalization, explorations are also made into its neural substrate where the general outlines are emerging but the unknowns are still vast.

The second or reciprocal algorithm (evolved rules of procedure or function) of behavior is set out from the tug and pull of ego and empathy. Three ranges in a spectrum of behavior are described: the egoistic, dominated by self-interested, self-preservation priorities; the empathetic, dominated by other-interested, affectional priorities; and the dynamic balance range, in which the priorities of ego and empathy are approximately balanced. These proposed algorithms, although individualized by genetic, gender, learning, and experiential differences, nevertheless constitute the neural architecture common to all human beings. Although they operate very imperfectly, these reciprocal algorithms allow us to get to reciprocity through conflict in our behavioral motives and actions. These algorithms are seen to have a shaping effect on all interpersonal behavior, from the simplest social interactions to the most complex. The proposed algorithms are statistical, like the second law of thermodynamics and quantum physics, in that they do not allow prediction of precise behavior at the basic unit of analysis, the individual, molecular, or subatomic level respectively...but only on the basis of statistical probability. They also function analogously to a quantum wave function.

Chapter 7 examines the two brain algorithms as manifested in various theories of psychodynamics. The algorithms, although not identified as such, not distinguished from each other, and largely implicit, are seen to fundamentally underpin the thinking of such writers as Friedrich Nietzsche, Sigmund Freud, Erich Fromm, Frederich Perls, and Jean Piaget.

Chapters 8-10 trace the expression of the two brain algorithms as the foundation of western and eastern philosophy and mysticism. The primary algorithm, especially, is seen to frame the process of reasoning in the west in the dialectical thought of the pre-Socratics and Plato, the syllogistic reasoning of Aristotle, the transcendental dialectic of Kant, the cosmic dialectic of Hegel, and the material dialectic of Marx. The dialectical thought of the west, following Plato took the form of a dialectic of becoming. On the other hand, the dialectic that developed in Chinese thought, originating in the yin-yang of the *I Ching* and in Taoism, expresses the brain algorithms from a different perspective. The dialectic of the Chinese lacks a forward movement of becoming characteristic of the west and may be thought of, instead, as a dialectic of being. The features of the primary algorithmic synthesizing loop come through clearly in both systems despite the fact that it is viewed from different perspectives. The second reciprocal algorithm of our evolved brain structure, the tug and pull of ego and empathy, combined with the primary algorithmic expression, is traced as the foundation of mystical experience and expression. The two algorithms, have never before been identified as such, nor have they been distinguished from each other. In the prior literature they have remained confused in both their philosophical and mystical manifestations.

Chapter 11 presents an interim summary of the discussion thus far and sets the tone for the remainder of the book. It examines the expression of the primary algorithm as manifested in current computer science as well as other scientific thinking. The digital computer is seen as the rigorous projection of the primary algorithm in skeletal form. The comments and descriptions of prominent scientists in the search for a grand unifying theory (GUT) are shown to express inadvertently the framework and process of the biosynthetic algorithm in thought process.

Chapters 12-14 explore the two brain algorithms in ethics and social theory through the writings of Martin Buber, Tielhard de Chardin, and Jurgen Habermas. These thinkers express mainly the second reciprocal algorithm of our evolved brain structure...the tug and pull of ego and empathy...which underlies the significant moral issues of humanity. The primary algorithm is also expressed, but incidentally rather than centrally. The two algorithms, although pervasively manifest, are not explicitly recognized, identified, or sourced by the three thinkers.

Chapter 15 examines the primary algorithm in contemporary post-analytic philosophy, chiefly through the writing of Richard Rorty. Without grasping the nature of the algorithm, Rorty, nevertheless struggles with its inevitable shaping effect on his own philosophical thought.

Chapters 16-20 explore the shaping effect of the two algorithms on human language through a discussion of the current state of linguistic theory. The primary algorithm is proposed to constitute the framework of the long sought universal syntax of language both from the standpoint of grammar and the lexicon. The second algorithmic matrix provides the somatic link to value or significance so necessary to meaningful social communication, experience, and to the unity of mind and body. The two algorithms, then, shape language in differing, but complementary ways.

The final chapters examine further the neural substrate of language, thought, and experience from the varying perspectives existing within the study of neuroscience and summarize the variant expressions of the primary and reciprocal brain algorithms as they have been manifested historically across the spectrum of our thought, behavior, subjective experience, and language. The expression of the primary algorithm is further traced in the theoretical literature of quantum physics.

PART ONE

BRAIN ALGORITHMS OF BEHAVIOR AND EXPERIENCE

Chapter 2

ALGORITHMS OF THE BRAIN

The miracle of the appropriateness of the language of mathematics for the formulation of the laws of physics is a wonderful gift which we neither understand nor deserve
(E.P. Wigner, Nobelist, physics, 1963) (Wigner 1960)

All significant experience of humankind is processed through the brain. This is true whatever the source of the experience. The source may be internal to the body, originating from the various bodily organs to include the brain itself, or it may be external, originating from the social and physical environments outside our bodies. The brain likewise processes the thoughts that arise from such experience and the language that expresses and communicates the thought and experience.

Since all significant experience, behavior, thought, and language from whatever source, is processed by the brain, it becomes of crucial importance to understand how the brain does this processing. We might ask such questions as: What does the brain do to what it processes? Does it merely store it in memory? Or does it assemble it in some way? Or does the brain even create experience, thought, and language out of its own structures and mechanisms? These questions are important because they bear on the very meaning of our lives, our sense of self, of identity, of the nature of the reality within which we function.

To answer these questions, we must of course, look to the brain itself. There are many ways to approach the study of the brain. We may look at the various brain parts and study their composition, their interrelationships, as well as their historical evolutionary development. We may also examine how the sensory perceptions of the environment are processed internally, as well as the products or outputs of the brain, to determine the characteristics, the features of their expression and organization.

In the chapters that follow, I will at times use all these approaches. To begin with, though, I will first lay the groundwork for understanding some of the processing characteristics of the brain. The brain processes behavior, experience, thought, and language by formulas or processes called ***algorithms***. Since the term algorithm is widely used in science, it is very important at the outset to understand what is meant by that term.

The term was derived from the name of a Persian mathematician of the ninth century, Mohammed al-Khowarizmi. He gets the credit for giving us step-by-step rules for the familiar arithmetic of adding, subtracting, multiplying, and division. His name got changed a bit since it came down to us in translation. So that al-Khowarizmi would probably not recognize his own name if he saw it today.

ALGORITHMS IN SCIENCE AND DAILY LIVING

So algorithm, then, came originally to mean a way or method of solving math problems, usually by breaking the problems down into simpler, sequential steps. Its usage, however, has now been extended. In computer science it is used to describe the various specific step by step procedures for solving problems so that they can be programmed into computers.

Some writers, however, have taken the word even further. For example, David Harel, of the Weizmann Institute of Science in Israel, sees algorithms applying to human decision-making in most areas of life...not only to business, science, and technology, but also to such mundane things as looking up a telephone number, knitting, filling out a form, and even cooking. In fact, algorithms have been compared to recipes, which provide a step by step process for making dishes for the family dinner.[1]

Indeed, *Algorithmics*, the study of algorithms, because of its wide applications, has been defined as a discipline distinct from but closely related to math and computer science. An international journal has been established (1986) to deal with the application of algorithms to computer science. It is called *Algorithmica*. Algorithms related to cooking, however, are still dealt with under recipes and are still to be found unimpressively in cookbooks.

EXAMPLES OF ALGORITHMS

Taken generically, algorithms can be expressed in a variety of ways. For example, a recipe or algorithm for cooking brown rice is very simple. Written in plain language, it goes as follows:

[1]Harel (1987: 5-13). Harel provides the most readable introduction to algorithmics in this work devoted entirely to the subject. Also see Wheatley and Unwin (1972). Other introductory texts are aimed at the more specialized reader, e.g., Brassard and Bratley (1988).

In a covered 2 and 1/2 quart saucepan, place 2 and 1/4 cups water;
Bring to boil
Stir in 1 cup brown rice
Bring back to boil
Turn heat to low and cover
Simmer 45-50 minutes or until all water is absorbed and rice is tender
Fluff with fork.

Or take a simple act of looking up a number in the telephone book. Suppose we want to find the number of our friend Brad Smith. We follow the familiar algorithm, we probably learned as children, automatically, without thinking.

Get a local area phone book
Turn to the residential listings
Go to the page where the S listings begin
Go down the S listings to the SMs
Go down the SM listing to the SMIs
Go down the SMIs to the SMITs
Go down the SMITs to the SMITHs
Go down the SMITHS to the SMITH Bs....
Stop at the listing for SMITH, BRAD
Jot down the 7-digit number that appears to the right.

From the familiar ordinary-language algorithms of cooking and looking up phone numbers, we go to more specialized expression of algorithms used in computer science.

Algorithms can be expressed both functionally and structurally. Computer algorithms can be expressed functionally in the form of flow charts.

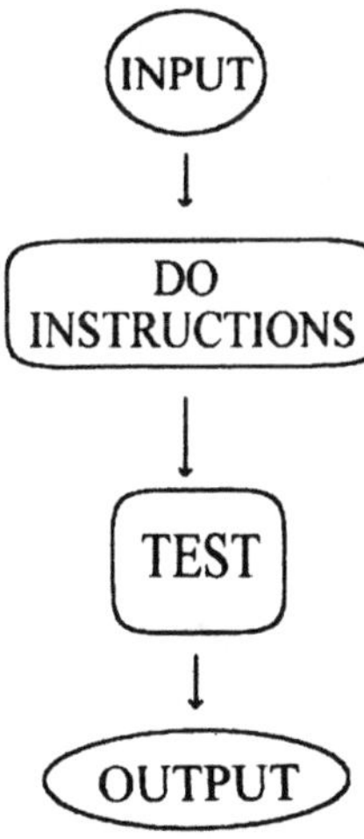

Figure 2-1. Algorithmic flow chart.

Flow charts show us graphically how the process or actions move in the algorithm. The inputs or raw materials are gathered as represented at the top of the chart...much as vegetables are gathered to put into a juicer. The inputs are then put into the processor where they're acted upon by a set of instructions...the vegetables are ground up in the juicer. The results are tested to see if they meet the criteria or our expectations...we sample the juice. If the processed materials meet the test, they are accepted as outputs. That is, we drink the juice and even serve it to our family and friends.

HIERARCHICAL TREES

Perhaps the most widely used and important way of representing algorithmic structure is the hierarchical tree. Trees are familiar to us in organizational charts, genealogical family trees, and schematics of almost anything. The simplest form of tree is as follows:

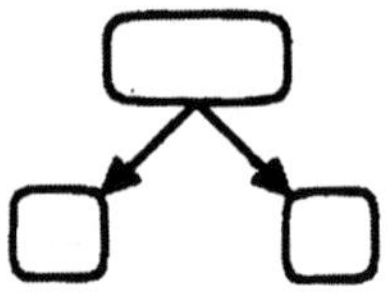

Figure 2-2. Upside-down hierarchical binary tree.

Figure 2-2 shows a typical upside-down tree as used in computer science. The so-called root is at the top with the branching proceeding downward. Since this tree has only two nodes, or offspring, as computer programmers say, it is known as a binary (two-part) tree. Trees can, of course, have as many nodes or offspring, as needed or can be reasonably handled.

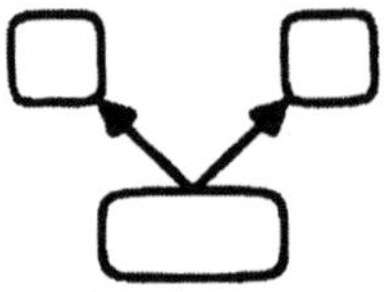

Figure 2-3. Right-side up hierarchical binary tree.

The upside-down binary tree can, if one chose, be flipped vertically to have its root at the bottom and its branching going upwards as are our more familiar botanical trees in nature (figure 2-3).

Trees, as noted, need not be limited to two nodes. They can have an infinite number of branches and nodes. Essentially, what they represent is a method of arranging parts into wholes...or if you like to think the other way...of breaking wholes down into parts.

According to Harel, trees with their numerous and varied applications...

> more than any other structural method for combining parts into a whole...can be said to represent the special kind of structuring that abounds in algorithmics (Harel 1987: 43-44).

Algorithmics, or the study of algorithms, then, is about the systematic study of combining parts into wholes in a rigorous way.

ARTIFICIAL INTELLIGENCE, COGNITIVE SCIENCE, AND NEURAL NETWORKS

The study of artificial intelligence, or cognitive science, attempts to produce algorithms that model or represent the processes of human thought. This is a very difficult and challenging task that despite advances sometimes seems frustratingly well beyond our present reach. The new study of artificial neural networks tries not only to represent the processes of thought, but also to make the structures of its artificial creations match as closely as possible the actual biological neural apparatus.[2] In other words neural networks aim at modeling algorithmic structures of the brain itself.

NATURAL ALGORITHMS: OF LIFE AND SOCIETY

There are, then, algorithms of behavior, experience and thought grounded in the nature of life itself and replicated in the human brain structure that evolution took millions of years of to produce. Such bioneurological, or natural algorithms run the basic mechanisms of our lives at all levels. Expressed also at a high level of generalization, they, likewise, shape our self-conscious individual and social lives. These algorithms have been expressed, occasionally explicitly, but largely implicitly, throughout the history of human thought and action. Since the same algorithms have been expressed in differing terminology in different areas of thought and study, the identity of the algorithms has gone largely unnoticed, unclarified. The

[2]There are a number of recent books devoted to neural networks. For example see Haykin (1994), Chester (1993), Karayiannis and Venetsanopoulos (1993), Churchland and Sejnowski. (1992), Levine (1991). Informative articles are also contained in *The Handbook of Brain Theory and Neural Networks* (1995); see especially the contributions of Barnden and Anderson.

failure to identify or distinguish these persistent, natural algorithmic processes has been the source of much illusion as well as confusion in the history of human thought and experience and has produced serious consequences in human social, economic, political, as well as spiritual life.

These natural algorithms of life, or brain structure, will be explored in their manifestations across the spectrum of our human thought, behavior, subjective experience, and language, chiefly at a macro- or high level of generalization. This is the level at which we experience ourselves consciously. Nevertheless, in the defining chapters (3-5) and toward the close of this book (chapters 21-23) I will make efforts to more concretely connect our thought, experience, and language with the mechanisms of the neural substrate.

Chapter 3

THE PRIMARY ALGORITHM OF OUR BRAIN: WHAT IS IT?

Curiosity demands that we ask questions, that we try to put things together....

(R. P. Feynman, Nobel prize, physics, 1965) (Feynman 1963: 1)

The search for the primary or fundamental algorithm or processing pattern of our brain leads us to examine the nature of life itself...to search out the answers to such questions as: What are we? How did we come about? Where are we going? These are the everlasting, nagging, yet always challenging questions of humankind.

Our modern science gives us a growing body of knowledge and many exciting insights into these questions. It also gives us one great barrier to understanding how it all fits together. This barrier is the division of science and academics into separate disciplines or domains...what biologist Lynn Margulis calls academic aparteid (1997; 1998: 175; also see; Dyson 1985:1-3; Schroedinger 1945: preface).

Instead of a science of life that gives us a unifying thread to help us make sense of things, we have a gaggle of partial "sciences" that fragment our sense of integration. We have physics, biology, psychology, linguistics, anthropology, sociology, economics, political science...with their arrays of subdisciplines. All these disciplines, however, must depend on each other. To be scientifically sound each must inevitably build on the other while adding its own unique perspectives and variables. That is, at some point, out of what we think of as the physics of non-living things, there emerges the biology of living things. Out of the biology of early living things evolves the amazing diversity of present life...and finally, humankind. And with humankind come the bio-social sciences of psychology, linguistics, anthropology, sociology, economics, et al., each adding its own variables and perspectives. And beyond all these, we have the more speculative, but essential human questions that constitute philosophy and theology.

Of course, in the absence of alien observers, even physics and biology didn't exist before we humans came along...since we are the only known life forms that create academic disciplines.

NOT REDUCTIONISM, BUT BI-DIRECTIONAL LINKS AND BRIDGES

Because each division of study adds its own variables and perspectives, sociology, and psychology cannot, then, be simplistically reduced to biology...and biology cannot be likewise simplistically reduced to physics.[3] Simplistic reductionism obscures the new and often unique contributions of each succeeding discipline. Reduction of the social sciences and psychology to principles of biology and physics would, therefore, give us a frustratingly crabbed, forced, and inaccurate view of our social and psychological lives.

On the other hand, however, when we go the other way...when we move out of physics and biology into the bio-social sciences, we cannot ignore the variables of physics and biology or assume them mistakenly into nonexistence. The variables continue to operate despite the shift in perspective that accompanies the new disciplines. The bio-social sciences depend totally on the human organism which is their subject matter. Without the human organism none of them would have meaning or even existence.

Recognizing the necessary relationships among the sciences, then, I begin the explorations into the fundamental algorithm of our brains, in effect, of our lives, at the beginning...or at least close to the beginning, with laws of physics. I will then proceed to, establish the linkages and cross the bridges into biology, psychology, language, and so on. In crossing the bridges and making the links, I will expose the often hidden algorithmic patterns that wind through and persist, to give the exploration meaning and unity...and, perhaps, a better understanding of what and who we are. In this process, I will be doing a degree of both reductionism and its opposite, integration. I will avoid simplistic reduction by moving up the scale of generalization, establishing links and bridges, from physics to the social sciences. In this enterprise, I will be moving in the direction of higher level integration. However, the steps may be traversed in reverse, retracing the links and bridges from the social sciences back to physics. In this sense, I will be doing reduction. Such a procedure is a nearly unavoidable feature of

[3]For a recent discussion presenting varying viewpoints on the question of the limits of reductionism in biology, see the report of the Novartis Foundation Symposium 213 (Bock and Goode 1998).

reasoning. It is often called the problem of levels of generalization or analysis. When viewed as an either/or problem, this procedure haunts all scientific thought and discussion. When not understood, it is especially troublesome in interdisciplinary scientific thought and discussion. The problem is algorithmic in nature and I hope to make this fact clearer further on. The fallacies to be avoided are getting hung up on one level of generalization or analysis to the exclusion of others and/or the taking of extreme, simplistic positions on the questions of reductionism and integration as either/or propositions.

Our leading sociobiologist, Edward O. Wilson of Harvard, catches the essence of the algorithmic process when he writes that much of the history of biological science may be seen dynamically as a tension between unit and aggregate, reduction and holism. Considering an equilibrium in this tension as neither desirable nor possible, he concludes that:

> As large patterns emerge, ambitious hard-science reductionists set out to dissolve them with nonconforming new data. Conversely, whenever empirical researchers discover enough new nonconforming phenomena to create ***chaos***, ***synthesizers*** move in to restore ***order***. In tandem the two kinds of endeavors nudge the discipline forward (all emphases are mine) (Wilson 1993: 243).

TIME'S ARROW: OR WHERE ARE WE GOING?

In physics, the quest for a primary structuring or processing algorithm begins with the laws of thermodynamics. These well-established laws of classical physics are among the most basic and reliable of science. The laws were discovered and defined in the 19th century by people working with steam engines, refrigerators, and the like to account for the heat flows that caused such machines to do the work expected of them. Since the original work was done largely on heat or thermal flows, the laws got the name "thermodynamics." The laws were later found to hold true for all energy flow systems, and became called also the laws of energetics. These laws are fundamental to the understanding not only of machines but also of living things. They, therefore, form an important link between physics and the next level discipline of biology.

The First Law: Conservation of Energy.

The first law of thermodynamics states that energy can neither be created nor destroyed. It can, however, change form.

On the purely physical side, for example, a stone sitting on the edge of a cliff has a store of potential energy because of its mass and the force of gravity acting upon it. It would have this energy whether we were around to observe it or not. If the stone is pushed or rolls off the cliff, some of that potential energy becomes kinetic energy (energy of movement) as the stone falls to a lower level and comes to rest again. No energy is created or destroyed in the process because, in its resting place, the stone still has the same potential energy as before. We can prove it by finding another cliff to push it off.

In the case of living things, the light energy from the sun is made into chemical energy by plants. Animals eat the plants and convert the chemical energy into energy to run their body processes and create kinetic energy to move about. The energy is returned to the earth and its surroundings by the animals in the form of waste products in a different energy form.[4] In the process no energy is created or destroyed, but it does change form. The first law doesn't tell us anything about where energy is going, only that even though it may change form many times, the total amount of energy remains the same.

The Second Law of Thermodynamics

It takes the second law of thermodynamics to tell us where we are going. The second law defines the overall direction of change and is often referred to as time's arrow.[5] Two important features of the second law are: irreversibility and entropy. Irreversibility and entropy can be illustrated by some simple drawings:

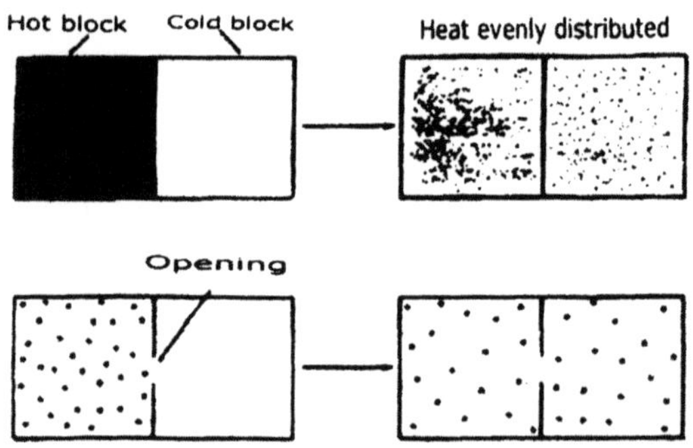

Figure 3-1. Diagram of hot/cold blocks and gas-filled/empty boxes.

[4]Sunlight is the overwhelming source of the free energy needed for life on earth. No other source could begin to provide the massive inputs of energy needed to sustain the living world as it exists and has evolved. See Asimov (1962, especially pp. 355-370) for one of the best discussions of solar radiation as the foundation for all life. Penrose (1991: 317-321) also provides a very readable discussion; also see Denbigh and Denbigh (1985: 118).

[5]For example, see Goldstein and Goldstein (1993: 7); Blum (1968). There is, also an excellent discussion in chapter 7, "Cosmology and the arrow of time," in Penrose (1991). Physicists have identified what seem to be three arrows of time, a cosmological arrow, a gravitational arrow, as well as the thermodynamic arrow. They appear to be linked in ways not yet fully understood. See the discussion in Davies and Gribben (1992: 112-139).

The above illustration shows crude drawings of the second law in action. The two systems, the hot and cold blocks (upper left) and the gas molecules (lower left), represent isolated (or closed) systems. Such systems are, by definition, systems that have no exchanges with their surrounding environment. Truly isolated systems are rare at best in reality but they serve a useful illustrative purpose in physics.

Take the first system. It represents a hot block of metal sealed together with a cold block of the same substance in an isolated system. The heat energy in the first block spontaneously flows toward the cold block until the temperature of both are equal. This state of equality is called thermodynamic equilibrium. The energy is evenly distributed throughout the system and is no longer available to do any work without outside help, like the application of more heat or cold. It is said by physicists to have reached a state of maximum entropy or disorder. Entropy can be a slippery term to catch on to but it means basically disorder, or decline of available energy to do work. Maximum entropy is maximum disorder, or as some say, chaos.

The flow of heat to the colder areas of this system represents an irreversible process. That is, the heat will not by itself, without the addition of some other energy process, flow back into its original condition of one hot box and one cold box. The second law of thermodynamics, then, provides that energy flows are by nature irreversible, that is, they do not reverse by themselves. And they proceed in the direction of increasing entropy or disorder.

The second system in the illustration...the gas molecules...shows the same phenomena. A box full of gas and an empty one are sealed in an isolated system. There is an opening between them. The molecules of gas flow from the full box into the empty box until each has the same amount. Again, the process won't reverse without outside help. The gas won't spontaneously flow back and gather into the original box again. The process is irreversible, and again, a condition of maximum disorder, chaos, or entropy is the final state.

Irreversibility is illustrated by many other phenomena; e. g., the melting of an ice block into water, the dissolving of sugar in a teacup, the burning down of a cigarette. Martin and Inge Goldstein have reminded us that if we were to watch a motion picture showing any of these things happening in reverse, we would immediately know that the sequence displayed impossible events, and that we were running the film backwards (Goldstein and Goldstein 1993: 6-7).

The second law of thermodynamics applies throughout the universe as we know it.[6] The overall direction is irreversibly toward increasing entropy. Stated simply, according to the second law, time's arrow, or the direction of process is:

ORDER → CHAOS

BUT WHAT ABOUT US? ARE WE GOING TO CHAOS?

We living things are such ordered creatures...so complex in our structure. How can a universe which is heading toward chaos produce us? Aren't we an exception? Don't we violate the second law? The answer, which physicists grappled with for a long time in the late 19th and early 20th centuries, is ...NO![7]

We, living things, grab things around us that are ordered, large amounts of them, convert them into our particular order or structure, use the energy, and then return them to the environment in degraded, or entropic form. As living organisms we are not isolated systems, but rather open systems that constantly exchange energy and materials with our environment. We impose our temporary order, in what biologist Charlotte Avers (1989: 54) calls a thermodynamically uphill direction (cf. Smith and Wood 1992: 2-5, who use the term, thermodynamically favored) by energy inputs from the environment. But we achieve this uphill structuring or ordering at the cost of increased entropy or disorder in the environment around us.

As philosopher and mathematician Alfred North Whitehead observed, living things are inevitable robbers of the environment (1929: 160). They have no choice but to take materials from their surroundings to keep up their life processes. Nobel prize-winning physicist Erwin Schroedinger building on concepts of his predecessor, Ludwig Boltzmann, put it differently by noting that living organisms sustain their order by constantly drawing negative entropy (order) from the environment.[8]

[6]Despite some uncertainty and disagreement over the applicability of the second law to the universe as a whole, the principle of entropy operates dependably within the biological timeframe. Morowitz writes that the first and second laws of thermodynamics are reflected in every aspect of metabolic organization (1978: 254). See also Lowenstein (1999: 3-17).

[7]Although the second law of thermodynamics is generally held to apply throughout the universe as a whole (Lehninger 1971: 26, Blum 1968), the recent work by Goldstein and Goldstein (1993) deals with the modifications of energy and entropy made necessary by the theory of relativity and quantum mechanics. Their conclusion is that although the second law must be modified, there is no good reason to doubt the continuing increase of entropy throughout the universe and the applicability of the second law as time's arrow. Prigogine writes that the arrow of time is the basic unifying element of nature (1994: 13).

[8]Schroedinger (1944). Schroedinger's short book, *What is Life?*, was seminal in discussions of life from the viewpoint of physics and information theory. See also Morowitz (1985: 220); Broda (1975: 2-3).

Bernard Rensch, a German biologist and philosopher, has called this in flowing terms, "the stream of order" whereby "order is continually determined by order."[9]

Living creatures, then, are ordering mechanisms. They take energy and materials from the environment, impose their own temporary order or structure upon them, suck out the usable energy in them, and then return the products to the environment in more chaotic form than when they took them in. They do this continuously, in a rhythmic, repetitive process governed by internal mechanisms,[10] until they cease to live.

The fundamental behavioral pattern of the living organism, from simplest algae right up the scale of complexity to humans is, therefore:

ORDER → CHAOS → ORDER →

Ad infinitum, or at least until the organism dies.

This living behavioral process can be thought of as a temporary intercept or holding action (e.g., see Williams and Frausto da Silva 1996: 281) against the relentless processes of the second law. But it is irreversible (we do not grow younger, but older) and it is done at the cost of increased entropy or disorder in the overall environment. There is no violation of the second law involved.

This algorithm, or process pattern, of ***ORDER →CHAOS →ORDER*** can be thought of as a bridge or link between non-life and life...a bridge between physics and biology. When we cross from the directly entropic (order to chaos) systems of physics into the ordering or self-organizing systems of living things, we are in the realm of biology.[11] In biology, new variables and perspectives of living behavior will be added, but they rest on the laws of physics. Those laws cannot be ignored or dismissed when we pass over the bridge.

---

[9]Rensch (1971: 77, 148-149). Rensch draws his position from Schroedinger (see 1967, chapter 6).

[10]For example, Nagai and Nakagawa(1992) report research findings that the suprachiasmic nucleus of the hypothalamus in rats is the neural substrate for the circadian rhythm that regulates feeding behavior and energy metabolism. By inference of homology this may also apply to other mammals including humans. Prior to the evolution of the nervous system such internal regulatory mechanisms were controlled by protein circuitry (e. g., see Bray 1995).

[11]The work by Prigogine and his associates on dissipative, or far from equilibrium systems, leads to the concept of order emerging at the edge of chaos (see Prigogine and Stengers 1984; Nicolis and Prigogine 1977; Prigogine 1980; Prigogine and Allen 1982; Prigogine 1983).

OUR BRAIN AS AN ORDERING MACHINE

The evolution of our complex brain makes all our science possible. We are here not only to order and use physical energy for our living processes, but also to observe, discriminate, interpret, order, and even alter the physical processes that pre-existed us. Our brain is a part of and affects the environment of our physical world. It evolved as our subject/object interface with the environment. This is an important point because it tends to be contested by some.

For example, Ramon y Cajal, the famed Spanish neuroscientist and Nobel Prize winner, reportedly observed that we can never understand our universe until we understand our human brain which creates that universe. And more recently, L. R. Vandervert suggests that our possible knowledge of the world is limited to what the brain can organize for us by its innate algorithms. He comments upon "science's erroneous placement of the world outside the skull"(Vandervert 1988, 1990: 9; see also Globus 1995; esp. p. 130).

Basically, such a perspective means that we can never reliably discover the true nature of the world because the brain will always impose its own algorithmic structure upon it and distort or filter out everything else. Such a perspective contains an aspect of truth. However, it can be overdone, leaving us with a false sense of futility and powerlessness.

The second view, recognized by most evolutionary biologists, is that the brain evolved as an interacting organic interface with the environment. It is not, therefore a stranger to the world, or a foreign product stuck into our environment from an alien universe. It is an adapted, functional part of the world in which it evolved. It's equally, if not more plausible, then, to argue that the brain provides reliable mapping algorithms or sets of algorithms that give us, for all practical (if not all scientific purposes), an accurate picture of the environment or world as it is...and as it is relevant to us. Not only our practical, daily lives, but also the great achievements of science like space travel, speak on behalf of accurate mapping.

There can be no arguing the fact, however, that the brain is also an ordering machine which ***can*** impose its structure on things we perceive. This is a very useful function and much of our science makes use of it. The position that I am maintaining here is that the brain's ordering patterns or algorithms for most purposes reflect an accurate mapping or picture of our environment. It may be true, however, that the ordering structures of the brain do not reflect reality at the level of microphysics. After all, our brain obviously did not evolve in, or to cope with, a microphysical environment. This may be a contributing reason why we perceive uncertainty in quantum physics and some other phenomena (e.g., see the discussion in Changeux and Connes 1995: 64-73; also Stapp (1972).

ALGORITHMS OF THOUGHT AND BEHAVIOR
WHY STUDY THEM?

Nevertheless, the algorithms of behavior, thought, subjective experience, and ultimately language, beginning with ***Order → Chaos →Order***, that I am identifying, may be viewed as either the algorithms imposed by our brain with no independent reality in the environment or else they can be viewed as our brain algorithms which for the most part match physical reality because they evolved adaptively to do so.

At any rate the algorithms pervade our reality in almost all disciplines and have done so throughout history. They have shaped our lives and continue to do so. Therein lies their importance and the sufficient reason for identifying and clarifying them.

BRAINWORKS

The brain can view and order things from many perspectives. For instance, take the construct of entropy just discussed. The brain not only creates the concept of entropy but it alters the concept to its own purposes. This can be seen in the concept's different uses.

Entropy, as used in thermodynamics, describes essentially an objective process. The process would happen whether we were here or not. It just wouldn't have a name. Heat passes from a hotter object to a cooler object until they both reach the same temperature and are therefore in a state of lesser order. And some heat escapes (but is not lost) to the environment in the process. It is an objective process, clearly repeatable, never reversible without the application of additional energy. Since we are here, we can see it with our eyes, feel it with our hands, measure it with our instruments, and conceptualize it with our brains.

The entropy of thermodynamics is not only objective, but at the molecular level it is statistical. The molecules hop around in such a seemingly random, haphazard way that it's impossible given the limitations of our instruments and the interference by the act of observation, for us to predict the specific movement of any one molecule. And some reversible states are statistically possible if improbable. But at the macro level the sum of the movements of all the molecules invariably follows the second law. Therefore the second law rests ultimately on statistical probability, partly perhaps, because we lack the instrumentation and capability if not the desire to define it purely deterministically by the laws of cause and effect at the molecular level.[12] All the foregoing would happen even if we weren't here.

[12]See the article on dynamical and statistical laws by Max Planck (1987), founder of quantum theory in physics and Nobel recipient (1918). In this well-known article Planck presents a lucid contrasting of dynamical laws which are directly causal (like the first law of thermodynamics) and statistical laws which rest upon probability (like the second law of thermodynamics). See also Goldstein and Goldstein (1993: 7). There is a tendency now to

In information theory, however, the perspective is different. It has often been asserted that in information theory, the second law and the degree of entropy and order associated with it, are not objective and statistical, but rather, ***subjective*** and statistical. This controversy arises because of the way we sometimes use the term entropy in information theory...to indicate our lack of knowledge about something...which may or may not tell us anything about the actual state of it in reality.[13]

In other words, in information theory, we may say something is chaotic, random, or disordered, just because we haven't yet figured out the order in it. In such a case we are using the term chaos as a measure of our ignorance. When we overcome our ignorance, then we'll say the thing is no longer chaotic but ordered.

That's the way our brain works: It works either to ***find*** order in what it perceives, or failing that, to ***impose*** order upon what it perceives.[14] Not only is our living body, then, an ordering machine, our brain (mind), is also an ordering machine. Many thinkers have described the mind as tending to make order out of chaos.[15] Evolutionary thinkers also describe the evolutionary process as the gradual tendency to greater order, greater complexity.[16]

---

treat both laws probabilistically or statistically, especially at the quantum level (see also Blumenfeld and Tikhonov 1994). Murray Gell-Man, theoretical physicist and Nobel recipient (1969), sees unpredictability and uncontrollability stemming from four sources; the fundamental indeterminacies of quantum mechanics, the phenomenon of chaos, the limitations of our senses and instruments, and our inadequate understanding and capacity to calculate (1994: 276).

[13]See the discussion in Denbigh and Denbigh (1985). The authors conclude that whereas information theory is based on a subjective interpretation of probability which represents our state of knowledge or ignorance, it is justifiable in the physico-chemical context to go further in regarding the probability distribution objectively as reflecting the actual nature of the system and the related physical constraints (1985: 117-118).

[14]Cognitive neuroscientist, Michael Gazzaniga, based upon split brain research, reports finding a device in the left hemisphere, which he calls the *interpreter*, which is always looking for order and imposing it even when there is none (1998: xiii, 156-159). Gell-Mann sees the tendency of the brain to impose order where there is none as a source of superstition (1994: 275-277).

[15]See the numerous articles on order contained in Kuntz (1968),which report the results of a seminar of leading scholars at Grinnell College.

[16]Evolutionary biologists recognize that the evolution of life has been toward increasing levels of complexity. Some have seen it as involving chance mutations acted on by natural selection. Others see mutations as not totally a factor of chance (driven by other causal factors), but random as to environmental adaptability and survivability. A group of more recent scholars see evolution as constructionist, based on self-organizing properties that produce material for natural selection to work on (Ho and Saunders 1984, Fox 1988,

OUR BEHAVIORAL FOCUS: THE SELF-REPLICATING, SELF-ORGANIZING ENVIRONMENTALLY INTERACTIVE CYBERNETIC LOOP

From this point on I will focus not on the algorithm of the ordering process of evolution --from simple organisms to more complex. Rather I will concentrate on the algorithmic process in our daily behavior as human organisms. This process of ***Order → Chaos → Order*** is the fundamental behavioral algorithm of life. It is the behavioral algorithm of protein-nucleic acid synthesis as it acts upon and within the environment. There is, further, historical precedent for seeing protein synthesis as the primary algorithm of life. The very term protein is derived from the Greek word "proteis", which means *first, fundamental, of primary importance*. The name was assigned in 1838 by the Dutch chemist, Gerardus Johannes Mulder (1802-1880) as he explored the structure of these basic substances (e.g., see Asimov 1989: 302; Axley 1998: 1).

Although it may be simply stated in informal terms, the behavioral algorithm should not be mistaken for a contentless high-level generalization. The structuring algorithm is a very complex and highly specified molecular biological process, involving almost incredibly intricate interaction between proteins and nucleic acids (RNA and DNA). Molecular biophysicist Werner Lowenstein (1999) has traced in detail the structuring algorithm, which may also be thought of as an information-processing or cellular communications process by which behaviorally active (environmentally interactive) three-dimensional proteins are generated from largely two dimensional nucleic acids and vice versa in a self-organizing, self-replicating circular, cybernetic loop.

The ordering algorithm, then, may also be properly defined in terms of information theory as a cybernetic process expressing the energy/matter and space/time dualities, based upon criteria set-points including branching and growth points, controlled by a combination of inhibitory and positive feedback mechanisms. A simple diagram from Lowenstein illustrates the essential nature of the looping or circular process.[17]

Kauffman 1993). Consult Gayon (1990), Grene (1990), and Brandon (1985) for a critique of the new developments and current controversy in evolutionary thinking.

[17] Although cybernetics or information control science may or may not map protein-nucleic acid structures directly and may or may not lead us to the actual substrates that perform the indicated processing, evaluating, and control functions (Changeux makes this same point about the relationship of mathematics to biology in Changeux and Connes 1995:60; see also the discussion following Quinn 1998: 126-127), Lowenstein's recent work (1999) certainly represents an impressive advance in this effort. See also Bray (1995). Neurophysicist Baev (1998) also applies the concepts of cybernetic control systems to biological neural networks. Compatible with Lowenstein and the views expressed in this chapter Baev sees such systems as expressions of protein control and regulation(1998: 208-211). See also Schmajuk(1997) for

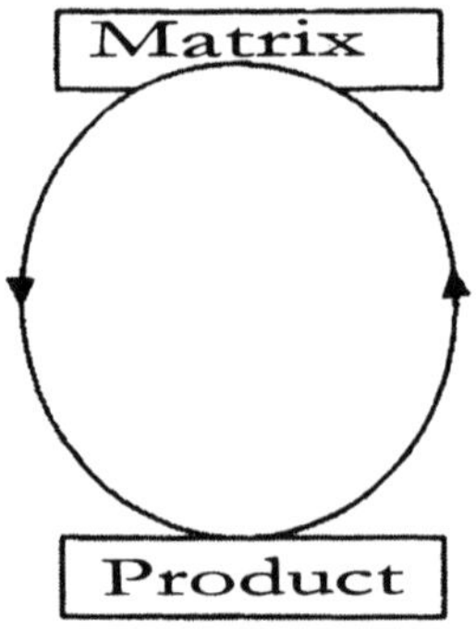

Figure 3-2. A self-promoting molecular information loop (after Lowenstein 1999: 33 with permission).

The detailed study of the biophysical-chemistry of the algorithm, in its various manifestations, at the molecular and cellular levels is the province of molecular biology. At the level of the brain and nervous system, it is the study of molecular neurobiology and the reader is referred to the works by Lowenstein and such molecular neurobiologists as C. U. M. Smith (1996) for detailed presentation. Both fields, although rooted in the past, have undergone explosive growth in the most recent decades. Molecular biology resulted from the convergence of biophysics, which concerned, primarily, the molecular structure of proteins and nucleic acids, and microbial genetics, which investigated the nature of heredity and the genetic process. The dramatic, accelerated development of the field dates from James Watson and Francis Crick's landmark solution to the structure of DNA in 1953. The applied outcome of these events led to the new high tech industry of biotechnology.

This highly varied, intricate, specified, and complex biochemical and biomolecular process of protein-nucleic acid synthesis with its related metabolic activity, in effect, defines and specifies the dialectical process of exchange between the organism and its environment. It is common to all life...and, of course, we share it.

This organic algorithmic process, with its essential cybernetic information-processing features, is replicated as an algorithm of our brain, our cognitive functioning. This is not surprising since the brain evolved, by modification and elaboration of the same molecular, cellular processes, in service to the organism...as the organism's increasingly complex and self-conscious subject/object interface with the environment. At the cognitive

a book that uses abstract neural network modelling to account for general associative learning phenomena without referring or mapping to the actual construction of the nervous system.

level, as at the cellular level, the algorithm is specified, although perhaps not so tightly, by the survival requirements or needs of the organism.[18] That is, it is bound to its genetic informational or somatic matrix from which it develops. It is ***never*** divorced from somatic or bodily meaning related to survival…even at the highest cognitive levels. Thus, it assures the essential unity of mind and body. It ***never*** functions solely as a contentless syntax. This is a very important point that I will return to in later chapters.

As replicated in our brain structure, the algorithm is the fundamental calculus of our behavioral and thought processes...expressed throughout our various disciplines somewhat differently from their unique perspectives. But it is readily identifiable once exposed. It is the first or primary bioneurological or brain algorithm of our thought, behavior, our language, and our subjective experience

[18] In recognition of its expanded, flexible characteristics in cognition, the algorithm may be thought of as a ***creative*** loop after Harth (1993).

Chapter 4

THE PRIMARY ALGORITHM AND BEHAVIOR: LIFE, WHAT DOES IT DO?

Every plant and animal acts as an incorporating center which brings order out of environmental disorder
(E. W. Sinnott, Biologist 1950: 30-31)

From the standpoint of physics, life orders environmental chaos. But it doesn't just build a one-time structure and stop with that, like inorganic crystals for instance. Life draws materials and energy from the environment in a continuous process to maintain its structure and fuel its behavior. As noted in chapter 3, when we pass from physics into the study of living things, we enter the academic realm of biology.[19] In doing so, we run into a vocabulary change as well as a change in perspective. The focus is now upon life processes; how they originated, and how they work. Biologically speaking, how can we tell something is alive? At what point in the history of earth do we recognize that nonliving things have become living?

PROTEIN SYNTHESIS AND REPLICATION

Recent discoveries of unquestionable archaic organic microfossils in Western Australia and South Africa date the appearance of life at about 3.5 billion years ago (Schopf 1993, 1996; Strother 1992). At that time the earth was already almost a billion years old and the universe was perhaps 16 billion years old. Although life appeared earlier than previously thought, it had to await the right conditions (e.g., see Prigogine and Stengers 1984:175-176). And there was a long period of chemical molecular evolution before

[19]For discussions of the laws of thermodynamics (energetics) as they apply to organisms, see Lowenstein (1999), Morowitz (1978), Broda (1975), and Lehninger (1971).

we finally saw something that was indisputably alive (Lowenstein 1999: esp. the graph on p.35; cf.; Fox 1988: the graph on p. 158; Williams and Frausto da Silva 1996: ch. 7-8; Dyson 1985).[20]

To be indisputably alive, an organism must be able to do at least two fundamental things:

1. It must convert energy from the environment, to structure and maintain itself.
2. It must make a copy of itself.

These two functions encompass the cybernetic looping process of protein-nucleic acid synthesis (or alternatively, biosynthesis).[21] Protein synthesis entails primarily the synthesis or build up of proteins from amino acid building blocks under the guidance of the informational matrix of DNA via the transfer functions of RNA. Energy is provided by energy-carrying molecules like ATP (adenosine triphosphate). Replication basically involves the synthesis of nucleic acids, assisted in turn by proteins, to code and transcribe the genetic information for replication. The two processes, thus, operate conjointly in the protein-nucleic acid synthesizing loop.

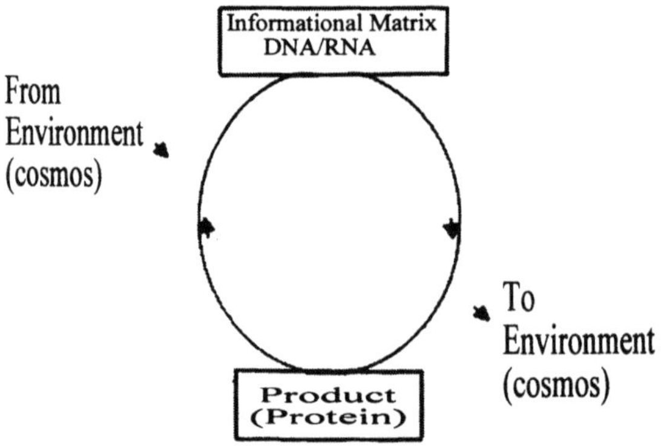

Figure 4-1. The protein-nucleic acid synthesizing loop, showing input/output to the environment in keeping with the second law of thermodynamics.

[20]Of course, in the standard model, cosmic evolution of elementary particles and basic atoms (hydrogen and helium) began shortly after the big bang between 10 and 20 billion years ago. Molecular evolution began on Earth about four billion years ago (Lowenstein 1999: 33-36).

[21]See Kauffman (1993: xv-xvii), Crawford (1972: 27-43), Broda (1975: esp. 30-35). For a somewhat novel discussion of the origins of metabolism and replication see Dyson (1985); also see Fox (1988).

Simple one-celled life forms like algae, as well as cyanobacteria and multicellular plants, follow the same synthesizing loop. They absorb or take their necessary energy from the sun in the form of photons and their necessary carbon from atmospheric carbon dioxide. Through a process long-called photosynthesis they store the sun's photon energy up in an informational molecular matrix form (nucleic acids), synthesize their essentially protein cellular structure, and, in turn, make another copy of themselves. Such a mode of nutrition, in which no prefabricated organic materials are used...that is, the organism makes it own organic materials...is called ***autotrophy***. In ***heterotrophy***, the nutritional form of most animals, prefabricated organic materials are required (Margulis, et al. 1994: 6-7). In keeping with the ***Order → Chaos →Order*** algorithm, photosynthesis is, of course, an ordering process. In fact, the word means ordering of light.

The basic behavioral function of an individual life form, if we use behavior to mean its actions as an organism toward its environment, is then, to absorb or incorporate materials and energy from the environment to maintain its structure, to run vital processes, and to replicate. Incorporation is a better word to use because it implies action, whereas absorption suggests a passive process. There is nothing passive about the intricate dynamics of protein-nucleic acid synthesis and its related metabolic activities. And incorporation is used widely in biology, psychology, and other biosciences. It means to take something into and make it a part of the body. A number of prominent biologists have described this incorporating behavior.

The Russian biologist Oparin makes the direct connection between the synthesizing metabolic process and incorporation when he writes:

> An organism can live and maintain itself only so long as it is continually exchanging material and energy with its environment... (which) are converted into the material of the organism itself. This is the ascending limb of metabolism known as assimilation (incorporation) (Oparin 1964: 9)

Edmund Sinnott, an American biologist, who taught for years at Yale University stated that:

> Every plant and animal acts as an incorporating center which brings order out of environmental disorder.[22]

And well-known cellular biologist Lynn Margulis and her coauthor, Dorion Sagan, write:

> The biological self incorporates not only food, water, and air—but facts, experiences, and sense impressions, which may become memories (1995: 32).

[22]Sinnott (1950: 30-31). Oparin is one of the earlier biologists who straightforwardly connects metabolism with incorporation (1964: 9). Others, of course, such as Sinnott, accept it implicitly. More recently Williams and Frausto da Silva have dealt with the mechanisms of incorporation in organic synthesis extensively (1996: see especially pp 393-416, 486-490).

The basic behavior, then, of the living organism vis-a-vis its environment is to order, or more descriptively, ***incorporate*** material and energy from the environment to build and maintain its structure and vital processes and ultimately to replicate itself. From the standpoint of evolution, the end purpose of this incorporation is replication ...because an organism that can't replicate can't evolve and would be only a passing phenomenon. From the perspective of the minute by minute, day by day activity of the individual living organism, however, incorporation is the primary behavioral activity or function. Elimination, or the return of matter and energy to the environment is a by-product. Replication is the ultimate end product. M. A. Crawford, writing on chemical evolution, noted that animals were the first life forms to eat other life forms. He traces the word "eat" back to the assimilation or absorption (both synonyms for incorporation) of by-products of the fermentation of photosynthesis, which later developed into the active transport of sugar and the digestion of proteins (Crawford 1972: 42-43).

Incorporation, or eating, as noted above, is the main behavioral thing the organism does. Incorporation is a still a better word because it implies, beyond eating, the converting of the ingested materials into the organism's own structure... the making of order out of relative chaos. The connection between eating and incorporation is an unavoidable one, however, and it reappears in psychological and especially mystical literature.

COMPLEXITY AND EXTENSIONALITY

Momentous as was the coming together of protein-nucleic acid synthesis to create truly living organisms, there was clearly a limit to what a single cell could do or become by just sitting there in a self-repeating cybernetic loop incorporating and finally replicating. Primitive life began early to evolve more complexity.

This complexity involved most conspicuously a reaching out into the environment to get the necessary energy and materials to incorporate. Some single celled organisms evolved hair-like appendages called cilia that let them move about in the water or soup to get what they needed. Others evolved whip-like parts, called flagella which whipped materials into contact with the main body of the cell where they could be incorporated. One-celled organisms like the ameba evolved pseudopodia (false feet), elongated parts of the cellular body, which they extended out into the environment to capture elusive materials and energy. The cilia of microbial organisms such as the gastraea, a form of plankton, did double duty for movement (extension) and the collecting of food particles for incorporation. The apical (top) area even had a primitive sensory organ to guide the cilial activity (e.g., see Nielsen 1995: 9-17).

As evolution progressed and animals became increasingly complex, they evolved fins, wings, arms, and legs to better move about and reach out to get what they needed. They also evolved sophisticated sensory organs to identify, target, and direct this reach into the environment. The amazing diversity of life on earth today is primarily an elaboration of this protein-nucleic acid synthetic process as it found ways to reach further and more effectively into the environment for materials and energy...always, of course, with the ultimate biological aim of replicating itself.[23]

Behaviorally speaking, this reaching out, this extension into the environment can be thought of as an evolutionary outgrowth of the basic behavior of incorporation...the accumulation of more and more interconnected incorporative, synthesizing cybernetic loops. Perhaps reaching out can be thought of as implicit in incorporation. The materials had to be got from somewhere. These two functions of incorporation and extension describe well the basic behavioral algorithm of protein-nucleic acid synthesis. In fact, they may be called the means of behavior. We can say that living organisms maintain and actualize themselves behaviorally by the means of extension into and incorporation of the environment. From the perspective of chemistry, R. J. P. Williams and J. J. R. Frausto da Silva, in their jointly authored work on chemical evolution, deal with the extension or outreach aspect of incorporation in their detailed discussion and analysis of ***uptake*** and ***incorporation*** of elements in living systems (1996: 393-416, 486-490). Molecular neurobiologist C.U.M. Smith makes the same point when he writes:

> It is possible to argue that the nervous system developed to serve the senses. Heterotrophic forms such as animals necessarily have to seek out their nutrients (1996: 2).

This seeking out of nutriments is captured neatly in the phrase ***motile heterotrophy*** (which means moving around to get nutrients) (Margulis, et al. 1994: 6). Animals show this seeking out, reaching out, or extending behavior most clearly, but plants, also, display these characteristics. Plants send roots into the earth, grow toward the sun, spread leaves to catch sunlight, and even do limited orienting movements called tropisms...all to catch matter and energy, or nutriment, to incorporate into themselves. Even fungi, which are often said to ***absorb*** nutriment, secrete (a form of reaching

[23]This functional development has long been recognized by biologists. See the classic work on lower organisms in Jennings (1906). More recent works are: Buchsbaum and Milne (1960), Russell-Hunter (1969), Fenchel (1987), Nielson (1995). The behavioral functions of extension and incorporation (although not called by those names) are clearly discernible in the development of respiratory and digestive organs (incorporation) and in organs of perception and locomotion (extension). See the article by A. S. Romer (1958). Barnes, et al. (1993) take a self-consciously functional approach which emphasizes the fundamental character of movement and feeding. The functions are elaborated by the process of mutation and natural selection, or by self- organization and selection (as some very recent scholars would have it).

out, or extension) enzymes into the environment to breakdown food particles into molecules which are, then, ***incorporated*** into the body and used to generate energy and build the protein cellular structures (Margulis and Schwartz 1988: 154). Taxinomically speaking, one of the main differences between plants and animals has traditionally been how they display the function of extension. Plants do it in a much more limited fashion than animals, which begin to move about freely in their search for matter and energy.

From biology, then, we get a living algorithmic process of:

INCORPORATION → EXTENSION → INCORPORATION →

Since an organism must continually reach out and incorporate matter and energy from the environment to sustain its structure and vital processes, this process is a rhythmic, almost pulsing, pattern governed by the organism's nucleic acid prescribed-protein executed internal mechanisms (e.g., see for higher organisms, see Nagai and Nakagawa 1992) that continues until death. It is the behavioral algorithm of protein-nucleic acid synthetic algorithm and its associated metabolic activity.

We now have two behavioral algorithmic patterns: one from physics, one from biology.

ORDER → CHAOS → ORDER

and

INCORPORATION → EXTENSION → INCORPORATION

The patterns are admittedly, in part, analogies. But they are more than that. A close look reveals that they are really expressive of the same algorithm albeit from different perspectives. The first pattern, coming from the perspective of the second law of thermodynamics, sees living things as ordering chaos. The second, from the viewpoint of biology, sees the organism as reaching out for environmental chaos so that it may order or incorporate it.

So that words don't fool us, we should note that ***incorporation*** of the second pattern is a virtual synonym for ***order*** of the first pattern...a synonym with a slight difference, that is, of a behavioral biological perspective. The extension and chaos stages of the two processes can likewise be viewed identically. When the organism extends, it acknowledges the breaking down into chaos and the necessity to get more materials for the ordering process.

These two process patterns express the primary algorithm of life. They are the patterns that evolution works upon as organisms ultimately take disordered or chaotic materials from the environment and impose their own order or structure on them in the act of replication. They are also the behavioral patterns of each organism during its individual life cycle with or without reference to replication.

As later replicated in our evolved brain structure and expressed in our thought, behavior, subjective experience, and language, the ubiquity of these patterns of our primary bioneurological algorithm is astounding. We will meet up with the expressive variants time and again as we explore the linkages between disciplines. The algorithm will be expressed differently from different perspectives. That is, the words will be different, but the meaning will be at bottom the same. The algorithm will be recognizable. To achieve this recognizability it will be necessary throughout the varying topical chapters of this book to connect the more formal "hard" science terms of physics and biology with the terms of the "softer" social sciences, and even with the varying literary, descriptive terms used historically in the literature of thought and philosophy.

The jump from protein-nucleic acid synthesis to cognition may seem a bit of a stretch to those who are unaccustomed to thinking across multi-disciplinary boundaries. Nevertheless, one must confront the undeniable fact that the entire life process began and is currently maintained by the algorithm of protein-nucleic acid synthesis. Everything that follows builds upon and is dependent upon this fact. Before the appearance of the organic algorithm with its related metabolic activity, there was no living algorithm. There is still no alternative algorithm.

THE INEVITABLE MATRIX-PRODUCT LINKAGE AND THE MIND-BODY CONNECTION

To summarize: In the synthesizing loops of the primary algorithm, proteins are the doers, the behavioral executives. The nucleic acids, however, structure the formation and capabilities of the proteins. The matrix-product linkage of the algorithmic synthesizing loop is unbreakable. ***Something*** is ***not*** created out of ***nothing***. The organic algorithm ***requires*** an informational matrix with cybernetic features. The behavioral proteins are, thus, inescapably connected, limited, as well enabled by the nucleic acid informational matrix. Proteins represent the ***behaving***, ***moving***, ***doing*** expression of the essentially passive ***do-nothing***, but nevertheless prescriptive, nucleic acid informational matrix. All behavior, then, being protein mediated, is connected directly or indirectly to the genetic informational matrix. This may be thought of as the link between behavior (to include emotional and cognitive), soma or body, and ultimately genome. The linkage may become abstract and rarefied in our higher cognitive acts of thought and language but it, nevertheless, inevitably remains. This is a crucial, though often obscured fact, which establishes the relation between cognition, behavior, soma, and genome...or essentially, mind and body.

In further confirmation of this important fact, the three dimensional protein molecules...representing the portion of the synthesizing loop that interacts with the environment...have been shown to be involved in information processing, computation, and information storage in living cells lacking neurons or nervous systems (Bray 1995). At higher levels, long-term memory or learning itself has been shown to involve the modification of protein structure and creation of new proteins in our neural anatomy (Dubnau and Tully 1998; Kandel, et al. 1995: 671-676; Macphail 1993: 59-61) Experientially or behaviorally, there is nothing else to modify except protein structure and the process and products of its related metabolic activities. The nucleic acids, on the other hand, are kept largely immune from experiential or behavioral modifications by protein molecular configurational barriers (see Lowenstein 1999: 108-117).

BRAIN ALGORITHMS AND THEIR COMPUTER EXPRESSIONS

It may be useful at this point to recall for purposes of comparison the algorithmic expressions from chapter 2. The figures from chapter 2 are repeated below for convenient reference.

The flow chart, the functional expression of algorithms used in computer science, shows the essential cybernetic characteristics of the protein-nucleic acid synthesizing loop. There is an informational matrix in the form of directions for processing to include evaluative feedback and control informational mechanisms (e.g., standards, comparators) Such an algorithm, using inputs from the environment, systematically performs the ***matrix to product*** loop.

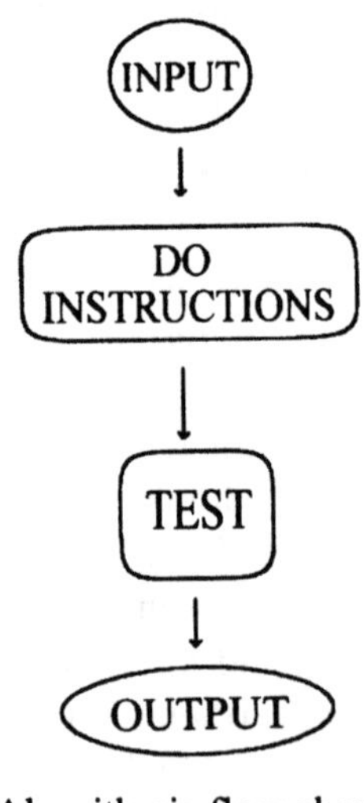

Figure 4-2. Algorithmic flow chart (figure 2-1).

The hierarchical tree…that widely used and important way of representing algorithmic structure…appears as a truncated expression of the organic algorithm in that the matrix may be implicit rather than directly represented. The ***steps*** toward the structuring of product are shown by the tree, but the inevitably necessary matrix may well exist implicitly in the programmer's head, in the program design, as well as in the hardware and operating system of the computer. Again, it is important to remember that something is not created out of nothing...the informational matrix is always present, explicitly or implicitly.

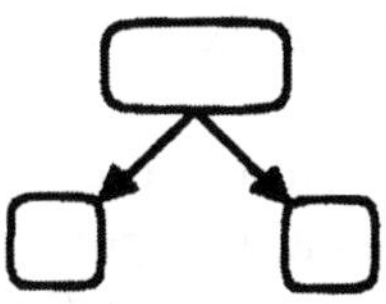

Figure 4-3. The hierarchical tree (figure 2-2).

VARIANT ALGORITHMIC TERMINOLOGY

It may be helpful to point out what is obvious, yet not so obvious. That is, the shift in vocabulary that occurs even in ordinary texts that obscures the elements of the primary algorithm. For example:

Whole* → *Parts* → *Whole (algorithmic tree, ch 2).

Order* → *Chaos* → *Order (physics, ch 3).

Incorporation* → *Extension* → *Incorporation (behavioral biology, ch 4).

Biosynthesis* → *Biodegradation* → *Biosynthesis (the two aspects of protein-nucleic acid synthesis, see any biochemistry text).

Anabolism (build up) → ***Catabolism*** (break down) → ***Anabolism*** (the two aspects of metabolism, see any biochemistry text).

Synthesis* → *Analysis* → *Synthesis (see any cognitive logic text).

A close inspection of the foregoing varying terminology quickly reveals the essential analogy, if not the algorithmic identity.

I should again emphasize that the protein-nucleic acid synthesizing loop, the primary organic algorithm, is the origin in physics and biology of the ***subject/object*** relationship that figures so pervasively in philosophy, psychology and other disciplines. The ordering (physics), incorporative (biology) behavior of the organism vis-a-vis its environment establishes it as ***subject*** which uses, exploits, its environment or parts thereof as ***objects***. Of course, this does not become self-conscious subject/objecthood until organisms develop a nervous system and brain with the matching bioneurological algorithm(s) capable of making such self-conscious discriminations or distinctions. Right now we think only humans have the level of brain development which permits this degree of self-consciousness.[24]

COSMOLOGY: OR THE QUESTION OF CAN SOMETHING COME OUT OF NOTHING?

Before leaving this chapter there is a pesky question, I feel compelled to deal with. That is, the question of the origin of the order...especially the organic order we see all around us. We've seen that the primary organic algorithm, the protein-nucleic acid synthesizing loop requires an information matrix in order for it to function in its ***matrix→ creates→ product→ creates→ matrix*** looping fashion. But where does the original information in the necessary informational matrix come from? After all, something cannot come out of nothing.

The ultimate answer...lying before the cosmic explosion that seemingly initiated our present universe...is at this point beyond the reach of science. From the Big Bang forward, however, to the origin of the informational matrix of the organic loop, science can piece together a pretty good explanation. I will summarize it briefly below.

Approximately 16 billion years ago, a super-dense, super-heated entity containing all the stuff of the universe exploded, throwing its energy outward in all directions. When the temperature cooled slightly, the four basic forces of physics...gravity, the strong nuclear force, the weak nuclear force, and the electromagnetic force...appeared. The first elementary particles also made their appearance. With the appearance of the four forces and the formation of basic particles we may arbitrarily say that what we call ***information*** made its appearance. I choose this point for the beginning of information because it is here that we first clearly have gradients or differentials of matter and energy. Although mathematician Norbert Wiener has been noted for saying that information is information and it is neither matter nor energy, information is, nevertheless, necessarily inseparable from matter and energy. Information is entirely dependent on differentials or gradients of matter and energy.

As time passed and the temperature from the original explosion cooled further, the elementary particles combined to form the simple nuclei of hydrogen and helium. Gaseous clouds of essentially hydrogen and helium, then condensed to form galaxies and the first generation of stars.

[24]In writing on the origins of the philosophical dialectic, Kainz (1988: 22-44) sees the subject/object distinction as the phenomenological experience most fundamental to the emergence of traditional logic. Following a distinction made by Hegel he sees the experience of self-consciousness as equally important with the subject/object experience in the origin of dialectics.

Information increased as the differentials and gradients of energy and matter became more pronounced and varied. The first generation of stars, upon consuming their hydrogen fuel, collapsed upon their cores creating the temperatures and pressures necessary to force the binding of nuclei and electrons into heavier elements. When these stars exploded into existing space as supernovas, they spewed out these newly formed heavier elements. From the condensation of these heavier elements, which represented ever more varied and complex gradients of energy and matter and thereby information, came our sun and other second generation stars.

On the planet Earth, formed about 4 billion years ago at the birth of the solar system, these newly created information-filled heavier elements (like oxygen, carbon, nitrogen, and the other heavys) combined with the earlier elements, particularly hydrogen, to gradually form the molecular components of the organic informational matrix. The matrix itself would evolve by essentially random or chance changes (mutations) in molecular structure which increased the informational content of the matrix and permitted or facilitated its survival and capacity to replicate itself.

The informational matrix of the primary organic algorithm, then, had its origin in the cosmic information-generating processes. It did not create something out of nothing, but drew upon these cosmic processes. All this was done in keeping with the second law of thermodynamics, which essentially governs or describes the cooling down of the universe from the intense temperatures of the Big Bang (cf. Brooks and Wiley 1988 for an interpretation of evolution as entropy or the thermodynamic cooling process).

Accompanying the entropic cooling process, we see the emergence of the four forces, the resulting creation of gradients of energy and matter, the accompanying increase of information, as well as the appearance and expansion of time and space. All these accompaniments are essential components of the primary algorithm itself and are reflected therein. Occasionally, I will refer back to them as I trace the many varied expressions of the organic algorithm in the chapters that follow.

Chapter 5

THE SECOND ALGORITHM: OUR BRAIN OF CONFLICT

If I am not for me, who will be? But if I am only for me, what (good) am I?
(Hillel) (Wigoder 1989: 341)

Our brain evolved out of accumulated mutations, confirmed by natural selection, in the genetic information repository of our DNA. RNA, the sister nucleic acid, continued its primarily intermediary role between the DNA repository and the translation of that genetic information into three-dimensional behaviorally-interactive proteins. The brain, then, develops in the individual under the prescriptive guidance of DNA which, despite a considerable amount of developmental and experiential plasticity, assures a high degree of fidelity in the replication of fundamental brain architecture. The main features of our brain, then, like the main features of our human body plan (head, trunk, arms, legs) are expressed in common across our human species. It is essential to remember that our brain structure is the three dimensional essentially protein representation of the largely two dimensional DNA informational matrix. Again, something is not made out of nothing. The connection between genetic informational matrix and phenotypic or individual behavior--the mind-body linkage--remains.

I have said earlier that the brain is an ordering mechanism. The fundamental algorithm of protein-nucleic acid synthesis, ***Order →Chaos →Order*** (***Incorporation → Extension → Incorporation***) is replicated in our brain structure as the primary algorithm of our thought, behavior, and subjective processes. It is common to all organisms. To get the second algorithm, however, we must go to our evolved human brain structure. This second algorithm appears later in the evolution of life and is ***not*** common to all life. It is a relatively latecomer upon the scene. Since our evolved brain structure holds the key to understanding the second algorithm, we must again look to the study of evolutionary biology, especially to the subdiscipline of evolutionary neuroscience.

Political scientist, Elliott White, in acknowledging the end of the empty organism perspective that had prevailed for the greater part of the 20th century in the form of behaviorist psychology and focusing on the importance of neurobiology as a necessary foundation for the sciences of human action, to include the science of politics, writes: "A science of human life that ignores the brain is akin to a study of the solar system that leaves out the sun."(1992:1). Primatologist Shirley Strum and social scientist Bruno Latour in their article "Redefining the Social Link from Baboons to Humans,"(1991), argue for a performative model of social interaction in which society is continually constructed or performed by active social beings.

More recently, cognitive scientist Steven Pinker of Massachussetts Institute of Technology, in *How the Mind Works* (1997), brings together the computational theory of the mind and the emerging discipline of evolutionary psychology with emphasis on information processing as the primary function of an evolved, adaptive modular brain. The concept of the brain, as a set of information-processing modules evolved independently to cope with specific adaptive problems as set out by Pinker and others, has become the standard for cognitive psychology. Cognitive science, combining the insights of evolutionary psychology, seeks to discover how the mind works by, in Pinker's words, "reverse engineering." That is, it identifies adaptive behaviors in the evolutionary environment and then back engineers them to postulate specific modules in the brain to deal specifically with the identified environmental challenges. Although care must be exercised to avoid the obvious easy-way-out fallacy of postulating a separate and dedicated brain module for each behavior, function, or emotion identified, the modular view is supported by research upon the brain itself. Cognitive neuroscience digs into the problem from the opposite side to discover the specific brain modules or neural structures that control these functions and behavioral responses.

The cognitive science approach, coupled with cognitive neuroscience and combined with evolutionary psychology, has progressively filled the empty organism and has given us the theoretical and empirical foundations for an active, performing organism. These approaches see not the largely blank slate brain of erstwhile behaviorism, but a brain chock full of interconnected modules designed for coping with environmental challenges.[25] Owing to the overemphasis on cognition as information processing, there has been a somewhat belated growth of a complementary literature that covers or seeks

[25]My position here is one of *weak modularity* (see Kosslyn and Koenig 1995: 44-47, 447) in contrast to the discrete, encapsulated modules proposed by Fodor (1983).

to include the neglected area of feelings, emotions, and the innate reward and response systems or modules that give pure cognition or information processing its subjective quality, value, or affective meaning (e.g., Panksepp 1998, LeDoux 1996, Restak 1994, Edelman 1992).

This book, by first defining the dynamic primary algorithm from physics and biology, carries on this theme of the active, performing organism. It can be seen as an effort to further clarify and define the performative, dynamic that proceeds from our evolved modular brain structure to shape our behavior, thought, subjective experience and language itself. It attempts further to integrate the findings of these new approaches with the earlier influential and more vintage insights of psychologist Abraham Maslow and neuroscientist Paul MacLean. The Maslow hierarchy of needs and the triune brain concept of MacLean have both been with us for a long time (Maslow's hierarchy for the better part of five decades, MacLean's for the better part of three) and are generally familiar.

MASLOW'S HIERARCHY

In Maslow's theoretical structure, needs are usually organized from bottom to top in the form of a staircase, or stepladder as follows: physiological needs (hunger, thirst), safety needs, belonging or social needs; esteem needs; and the self-actualizing need. Maslow theorized that these needs were emergent: That is, as we satisfied our basic needs of hunger and thirst, our safety needs would then emerge. As we satisfied our newly emerged safety needs, the next level, the belonging or social needs, would come into play. Next came esteem needs, and finally, as these were satisfied, the self-actualizing need at the top of the hierarchy emerged (Maslow 1943, 1968, 1970).

Maslow's hierarchy has appeared in every basic text on psychology and behavior for the past four decades. It also appears in most texts on organizational behavior. Its influence has been widespread as a behavioral scheme of ready and easy reference. It has also been popularized in casual and impressionistic writing about motivation. Maslow's well-known concept represents one of the earliest comprehensive efforts to develop a model of the human biological inheritance.

The Maslow hierarchy has, however, serious shortcomings that limit its utility for conceptualizing the genetic inheritance. For one thing, it lacks an evolutionary perspective. The hierarchy of needs is presented as a given, disconnected from the evolutionary process which produced it. Secondly, the concept of hierarchy is not fully developed. It does not allow sufficiently for interaction of the levels of hierarchy and does not account for those cases that violate the normal priority of needs (Cory 1974: 27-29, 85-86; Corning 1983: 167-72; Maddi 1989: 110-118; Smith 1991). Maslow's hierarchy has also been criticized for being culture bound, fitting neatly with particularly the U.S. concept of material achievement and success as a steady stairstep progression of higher development (Yankelovich 1981). It thereby tends to ignore or diminish the great accomplishments in thought, morality, and service to humanity of many of the great figures of human history (Maddi 1989).

MACLEAN'S MODULAR CONCEPT: MISMEASURED AND MISUNDERSTOOD

MacLean's triune brain concept is one of the earliest modular concepts of the brain. Although it has been acknowledged by some scholars to be the single most influential idea in brain science since WorldWar II (e.g., Durant in Harrington 1992: 268), it has largely been overlooked by cognitive psychology. In an extreme case it has been summarily and undeservedly rejected as wrong by Steven Pinker in his recent book noted earlier. This anomalous situation, in which the pioneering modular statement of brain organization coming from neuroscience itself and providing a natural match with aspects of the modular cognitive approach, has been brought about by a couple of seriously flawed reviews of MacLean's work that appeared in the influential journals *Science* (1990) and *American Scientist* (1992).

The effect of these faulty reviews has been to deny the use of MacLean's very significant research and insights to the researchers in the cognitive psychological as well as the social science community, who relied upon the authority of these prestigious journals. In fact Pinker bases his unfortunate and mistaken rejection of MacLean's thought solely on a reference to the review in *Science* which is the most prejudicial and grossly inaccurate of the two (Pinker 1997: 370, 580). The detailed and documented rebuttal of these reviews is reported in appendix 1, and has been previously reported by the author (Cory 1998, 1999). The presentation that follows here is adjusted to accommodate criticisms where valid.

THE INTERCONNECTED, THREE-LEVEL (TRIUNE) BRAIN

In a thoroughgoing, encyclopedic summary of the prior fifty years of brain research, MacLean (1990) documented the human brain as an evolved three-level interconnected structure. This structure comprises a self-preservational, maintenance component inherited from the stem reptiles of the Permian and Triassic periods, called the protoreptilian complex, a later modified and evolved mammalian affectional complex, and a most recently modified and elaborated higher cortex.

As brain evolution progressed in the line ancestral to humans, simple protoreptilian brain structure was not replaced, but provided the substructure and homologues for subsequent brain development while largely retaining its basic character and function.

Accordingly, the brain structure of early vertebrate life forms ancestral to humans (i.e., early fishes and reptiles) became the substructure and provided the homologues for the mammalian modifications and neocortical elaborations that followed and which have reached the greatest development in the brain of humankind. Appreciating the qualitative differences of the three levels is important to understanding the dynamics of human behavior.

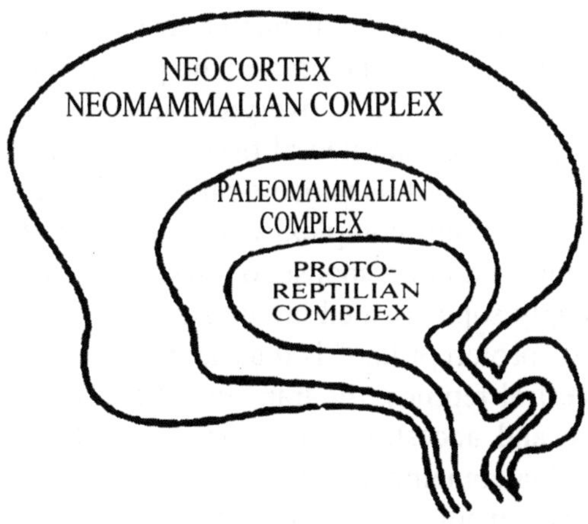

Figure 5-1. A simplified, modified sketch of the triune brain structure (After MacLean). As represented here the three brain divisions do not constitute distinct additions but rather modifications and elaborations of probable preexisting homologues reflecting phylogenetic continuity.

The protoreptilian brain tissues in humans are proposed, as they did in the ancestral stem reptiles, to govern the fundamentals, or the daily master routines, of our life-support operations: blood circulation, heartbeat, respiration, basic food-getting, reproduction, and defensive behaviors. Such functions and behaviors were the essential routines also to be found in the ancient stem reptiles. Located by MacLean in what are usually called the hindbrain and the midbrain (i.e., the brain stem) as well as in certain structures at the base of the forebrain, this primal and innermost core of the human brain makes up almost the entire brain in fishes and amphibians.

The next developmental stage of our brain, which comes from rudimentary mammalian life and which MacLean called the paleo- or "old" mammalian brain, is identified with the structures designated collectively as our limbic system. Developing from homologues preexisting in the protoreptilian brain, these newly elaborated limbic tissue-clusters included such physiological structures as the amygdala, the thalamus, the hypothalamus, the hippocampus, and other structures. Behavioral contributions to life from these modified and elaborated paleo-mammalian structures, or limbic system, included, among other things, the mammalian features (absent in the stem vertebrates) of warm-bloodedness, nursing, infant care, and extended social bonding. These new characteristics were then neurally integrated with the life-support functional and behavioral circuitry of the protoreptilian tissue complex to create the more complex life form of mammals.

The neocortex, which MacLean called the neo- or "new" mammalian brain, is the most recent stage of brain modification and development. This great mass of hemispherical brain matter that dominates the skull case of higher primates and man, by elaborating the preexisting homologues present in the brains of early vertebrates, overgrew and encased the earlier ("paleo") mammalian and protoreptilian neural tissues, but essentially did not replace them. As a consequence of this neocortical evolution and growth, those older brain parts evolved greater complexity and connectivity in support of these new tissue structures and in response to the behavioral adaptations necessary to life's increasingly sophisticated circumstances. The part of our brain that MacLean termed "protoreptilian" is thus actually considerably more complex than that of ancestral reptiles, the transitional therapsids of the Permian and Triassic periods, and ancestral mammals. And the part of our brain that MacLean called "paleomammalian" is also much more complex than that of ancestral and lower mammals. Each part, however, may be considered to serve, in the main, its original functions: our protoreptilian brain structures, though modified from those of ancestral reptiles, nevertheless principally regulate our basic survival mechanisms, while our paleomammalian structures, though modified from those of other mammals, principally govern our nurturing behaviors.

The unique features of the human brain evolved over a period of several million years in a primarily kinship based foraging society where sharing or reciprocity was essential to survival and which reinforced the adaptive evolution of the mammalian characteristics of self-preservation and affection.[26] Ego and empathy, self-interest and other-interest, are key features of our personal and social behavior. To relate these to MacLean's concept we need a subjective/ behavioral rather than a neurophysiological vocabulary...one that will express what the presence of our protoreptilian and

[26]For example see Humphrey (1976), Isaac (1978), and Tooby and DeVore (1987). Cosmides and Tooby surmise that cognitive development in humans allowed a widening and diversification of items of social exchange (1989: 59).

paleomammalian brain structures mean with regard to our day-to-day, subjectively experienced, behavioral initiatives and responses to one another and the world we live in. In computer-related vocabulary, familiar to us all through cognitive psychology and artificial intelligence, I use the software designer's vocabulary of programs and programming. I will speak of our three developmental brain levels as behavioral programs or sets of programs that ***subjectively*** drive and generate specific, and ***objectively*** observable, behaviors.[27]

From the predominantly survival-centered promptings of the ancestral protoreptilian tissues as elaborated in the human brain arises the motivational source for egoistic, surviving, self-interested subjective experience and behaviors. Here are the cold-blooded, seemingly passionless, single-minded behaviors that we've generally associated with the present-day lizard, the snake, and that most maligned of fishes, the shark.[28]

Here is a world revolving almost exclusively around matters of self-preservation. The protoreptilian brain structures, then, will be referred to as our self-preservation programming.

From the infant nursing, care-giving, and social bonding initiatives and responses of the mammalian modifications and elaborations arises the motivational source for nurturing, empathetic, other-interested experiences and behaviors.

Here are the warm-blooded, passionate, body-contacting, bonding behaviors that we've come to identify with the lion, the wolf, the primates.[29]

Here is a world in which nearly single-minded self-preservation is simultaneously complemented and counterpoised by the conflicting demands of affection.

[27]For earlier versions of the behavioral model developed here see Cory (1974, 1992, 1996). Also compare the model of human communication by Dingwall (1980) based in reflexive (striatal or reptilian) affective (limbic or paleomammalian), and cognitive (neocortical or neomammalian). Dingwall draws upon Lamendella (1977). See also Leven (1994).

[28]Experimental work in animals as diverse as lizards and monkeys shows the reptilian complex is involved in displays of agonistic and defensive social communication. Also it is noteworthy that partial destruction of the reptilian complex eliminates the aggressive, territorial display (MacLean 1993: 108).

[29]The division of function between the protoreptilian complex and the limbic system is not clear-cut, but rather entangled. The lower structures of the limbic node have been shown to augment the self-preservational behavior of feeding, fighting, and self-protection (MacLean 1990; 1993: 109), adding passion or emotion to them (Kandel, et al. 1995: 595-612). The newer structures in the upper half of the limbic node, especially the septal, including the medial preoptic area, and thalamocingulate division, are involved in the affectional, family-related behavior (Fleming, et al. 1996; MacLean 1993: 109).

The early mammalian modifications, then will be referred to as our affectional programming.[30]

Before I go on to discuss the neo-mammalian neocortical structures in behavioral terms, I wish to pause to consider how these first two sets of programs function together.

OUR EVOLVED BRAIN AND THE SOURCES OF SUBJECTIVE/BEHAVIORAL CONFLICT

These core behavioral program modules, composed of (or served by) sets or subsystems of modules of our brain structure, serve as dynamic factors of our behavior. They are energy-driven by our cellular as well as overall bodily processes of metabolism as mediated by ***hormonal, neuro-transmitter***, and ***neural architecture***. Each is an inextricable part of our makeup, because each is "wired into" our brain structure by the process of

[30]Positing the affectional programming draws not only upon current neuroscience but also the extensive literature on the concepts of social bonding and attachment, especially the work done on higher primates and man. For fundamental work on lower animals see the pioneering work of the Austrian ethologist and Nobel prize winner Konrad Lorenz (1970, 1971). Particularly relevant here would be the work of psychologist, Harry F. Harlow on the nature of love and attachment in rhesus/macaque monkeys (1965, 1986). Harlow described five affectional systems in monkeys--maternal, mother-infant, age-mate, heterosexual, and paternal (1986). Here I have proposed one all-inclusive affectional program. It is personally interesting to me that Crews (1997) argues that affiliative behaviors evolved from reproductive behaviors. This is a position I took in 1974 in the first version of the conflict systems neurobehavioral model (Cory 1974) presented in this chapter. There has been a recent resurgence of interest in the evolutionary biological basis of affection and empathy, especially in primates (e.g., Goodall 1986; de Waal 1996; deWaal and Aurelli 1997). In the case of humans, the work of Spitz (1965) and British psychiatrist John Bowlby (1969, 1988) are of special interest. All the foregoing reflect field observations, experimental behavioral observations and clinical work. None of them penetrate the brain itself. More recent work in computer modeling of neural processes has focused primarily on cognition and avoided dealing with the more complex issues of affiliation and emotion. For example, Churchland and Sejnowski in their extensive and well known work on the computational brain acknowledge the neglect of these critical areas (1992: 413). Some exceptions to this neglect include Levine and Leven (1992). From the standpoint of neuroscience, it is also notable that Kandel, Schwartz, and Jessell, authors of the most widely used text on introductory neuroscience also show this neglect (1995). Carter, et al. note that affiliation, as an independent topic, got little attention before 1990 (1997: xiii). Extensive research has been done on the role of the amygdala in emotion, but such research has generally focused on the emotion of fear (LeDoux 1997). The neglect is not difficult to explain. Research on such complex pathways within the brain, in spite of great progress in recent years, is still in its very early stages. The unknowns are still very vast. Currently the best summaries of research in neuroscience on the nurturing, caring, family-related behavior are contained in Carter, et al. (1997); Fleming, et al. (1996); MacLean (1990: 380-410, 520-62). In an effort to correct this imbalance in neuroscience Panksepp has recently applied the term *affective neuroscience* to his work which brings together research on affect and discusses the challenges facing it (1998).

evolution. The degree of genome control seems, however, to vary with the mechanism. Older brain parts like the hindbrain and parts of limbic system, phylogenetically old and necessary for survival, seem to be more strictly under control of the genetic informational matrix. Other more recent tissues in the neocortex appear more plastic and depend also on development and environmental experience. Damasio (1994) uses the term ***preorganized***, apparently (and appropriately, I think) to avoid the implication of an overly-deterministic prewiring in some brain regions. Behavioral conflict exists, then, simply by virtue of the presence of these two large-scale energy-driven modular program sets in our lives--up and running even prior to birth. Their mere physiological presence sets us up for a life of inner and outer struggle, as we are driven by and respond to their contending demands.[31]

Indeed, if you sensitize yourself to the dynamic striving for ascendancy that occurs between these two contending programs in your own nature, you may well be astonished by the frequency of the tug-and-pull going on between them. Much of our thinking, our internal monologue, is driven by this striving—largely below the threshold of awareness. And this tug-and-pull, this internal conflict, has been relentlessly and often blindly shaping our lives since we stepped forth into history.

CONFLICT: IT'S A FACT OF LIFE

The records of ancient history, hieroglyphics from the monuments of Egypt, cuneiform tablets from Mesopotamia, as well as the ancient texts of India and China, record and commemorate conflicts between ancient states and empires. The historical conflict between individuals, groups, tribes, and political entities continues today among modern nations. There has never been a society without conflict. Some have more, some have less, but none are without it.

And the central and indelible presence of conflict in human life has not been lost on our greatest thinkers or systems of thought…

> Heraclitus wrote, "The opposite is beneficial…all things are born through strife."
>
> Socrates saw human nature as made up of two winged steeds, one noble, one ignoble, harnessed to a single chariot and struggling against the control of the charioteer.
>
> Hillel, experiencing the tension between the two opposing promptings in his life, wrote, "If I am not for me, who will be? But if I am only for me, what (good) am I?"

[31]In cognitive neuroscience brain modules are commonly seen as competing and also cooperating (e.g., see Crick 1994; Baars 1997; Dennett 1998). The idea of competing or conflicting modules contriving behavioral tension is also acknowledged by Pinker (1997: 58, 65).

From Taoism to Buddhism, from Judaism to Christianity, our central moral themes have arisen from and examined the dualities posed by this tug and pull between preservation of self and affection for others.

Conflict, then, is in the nature of humankind, pervasive and inevitable. The eternal moral and ethical dilemma—Hillel's question: Do I serve myself or others?—is wired irrevocably into our human nature and carried with us right into almost every aspect of our daily lives.

Conflict is more than an externalized, objective ethical, moral, or decision-making dilemma, however. ***Subjectively, feelings of satisfaction*** occur when we can express our felt motives, while ***feelings of frustration*** occur when either our self-preservational or affectional impulses cannot be expressed in the behavioral initiatives and responses we wish to make.

Behavioral tension then arises. Experienced as subjectively defined variants such as frustration, anxiety, or anger, behavioral tension occurs whenever one of our two fundamental behavioral programs--self-preservation or affection--is activated but meets with some resistance or difficulty that prevents its satisfactory expression. This subjective tension becomes most paralyzing when both programs are activated and seek contending or incompatible responses within a single situation. Caught between "I want to" and "I can't"--e.g., "I want to help him/her, but I can't surrender my needs"--we agonize. Whether this tension arises through the thwarted expression of a single impulse or the simultaneous but mutually exclusive urgings of two contending impulses, whenever it remains unresolved or unmanaged, it leads to the worsening condition of behavioral stress.

THE BLESSING OF TENSION AND STRESS

The evolutionary process by which the two opposite promptings of self-preservation and affection were combined in us enhanced our ability to survive by binding us in social interaction and providing us with the widest range of behavioral responses to our environment.[32] Our inherently conflicting programs are a curse, then, only to the degree that we fail to recognize them as a blessing. Our self-preservation and affection programs allow us a highly advanced sensitivity to our environment, as well as the ability to perceive and appreciate the survival requirements of others. Ironically, the accompanying behavioral tension--even the stress!--is an

[32]The evolution of the neocortex, our big brain, was in all probability greatly enhanced by the tug and pull of our conflicting programs. Humphrey (1976) sees the function of the intellect providing the ability to cope with problems of interpersonal relationships. See also the discussion in Masters (1989: 16-26) and Erdal and Whiten (1996). Cummins (1998) argues that interpersonal relationships, competing and cooperating with conspecifics for limited resources, is the chief problem confronting social mammals. Cummins concentrates on dominance hierarchies which she sees as dynamic rather than static.

integral part of this useful function, for it allows us to more immediately evaluate (a subjective function) our behavior and the effect it is having on ourselves and others.[33]

Behavioral tension serves as an internal emotional compass that we can use to guide ourselves through the often complicated and treacherous pathways of interpersonal relations.

Behavioral stress tells us that we are exceeding safe limits for ourselves and others and for our larger social, economic, and political structures.

Behavioral tension and stress are, at this point perhaps needless to say, inherently and necessarily subjective.[34]

But of course all of this requires a certain level of consciousness, perhaps best designated self-aware consciousness, coupled with the ability to generalize our internally experienced motives. If all we possessed were the conflicting programs of self-preservation and affection, we would, of course, be among the life forms whose behaviors are governed by instinct. We would be driven by the urgings of fight, or flight, or bondedness; and every so often--like the legendary mule who, thirsty and hungry, looked back and forth between water and hay, unable to move--we would be caught in the indecision of those urgings.

But whether or not other mammals with paleomammalian brain structures, with self-preservation and affectional programming, experience conscious conflict from these two behavioral priorities, we certainly do. We can reflect and generalize, not only upon our choices, but also upon the meanings they have for our personal as well as our species' existence and significance. And it's in that capacity to reflect, to self-consciously experience, generalize, and decide upon the tug-and-pull of our conflicting urgings, that we come to third stage of brain development in MacLean's model: the neomammalian or "new" mammalian brain structures...what I have designated the executive programming.

[33]Damasio's somatic marker hypothesis by which emotions become connected by learning to certain behavioral scenarios is an example of a functional mechanism for producing behavioral tension/stress (1994: 165-201). Also see the comment on chronic mental stress (1994: 119-120). Tension and stress are mediated by hormones and neurotransmitters acting within neural architecture, rather than through the so-called hydraulic pressure model of earlier psychodynamic models.

[34] On the clinical side, behavioral tension and stress may be manifested in a variety of clinically defined disorders. Although it is beyond the scope of this book to consider such matters, it is appropriate to acknowledge the developing clinical literature which proceeds from a neuro-evolutionary-ethological perspective; for example, see Price (1967), Gardner (1982), Bailey (1987), Price and Sloman (1987), Wilson (1993), Price, et al. (1994), Allan and Gilbert (1997). Gardner and Cory (forthcoming) present a series of articles which lead in the direction of integrating clinical perspectives with social theory.

THE CONFLICT SYSTEMS NEUROBEHAVIORAL (CSN) MODEL

The neural substrate of consciousness is still a matter of considerable speculation and debate (e.g., see Tonomi and Edelman 1998; Searle 1997; Damasio 1994). Although the mechanisms are still unclear, I follow the position here that there is no homunculus (little person) or other Cartesian dualistic process involved. Nevertheless, it seems that our expanded and elaborated neocortex (or isocortex), anchored in and interconnected with our earlier mammalian and protoreptilian brain systems, is clearly part of the "dynamic core" (Tonomi and Edelman 1998; cf. Dennett 1998) necessary to our ***self-aware*** or ***self-reflective*** consciousness. As well, the neocortex provides us with the evolutionarily unique and powerful ability to use verbal and symbolic language to create concepts and ideas by which to interpret our consciousness...to describe the feelings, motives, and behaviors that arise within us and in response to our social and environmental experiences.[35]

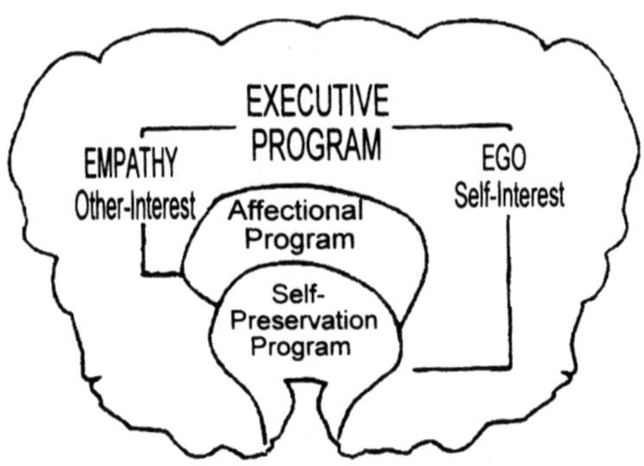

Figure 5-2. The conflict systems neurobehavioral model. A simplified cutaway representation of the brain showing the behavioral programs and the derivation of Ego/self-interested and Empathy/other-interested motives and behaviors.

[35]A language module did not, of course, pop out of nowhere and appear in the neocortex. The capacity for spoken language involved modifications of supporting anatomical structures including the laryngeal tract, tongue, velum (which can seal the nose from the mouth) and the neural connections that tied in with motor areas necessary for the production of speech. These all evolved relatively concommitantly from the hominid ancestral line and, combined with the elaboration of the neocortical structures of thought and syntax, made language possible. This example of the complexity of language development provides a caveat to avoid overly simplistic one for one specialized module for specific behavioral or functional adaptation positions. The work of Philip Lieberman, a linguistic psychologist at Brown University is especially relevant for the understanding of this very complex language capability. See the up-to-date treatment of these issues in Lieberman' s *Eve Spoke* (1998).

Although the positioning of Ego and Empathy in figure 5-2 (facing the reader) is primarily for illustrative purposes only and is not intended to indicate a definitive lateralization, there is some evidence to suggest that the right hemisphere is favored for emotion and the left for more analytical self-preserving behaviors (e.g., see Damasio 1994; Tucker, Luu, and Pribram 1995; Brownell and Martino 1998). However, Heller, et al. (1998), after noting that it is well established that particular regions of the right hemisphere are specialized to interpret and express emotional information, argue that the total experience of emotion is not lateralized but involves dynamic interactions between forward and posterior regions of both hemispheres as well as subcortical (limbic) structures. Such complex, highly generalized capacities as ego and empathy should more safely be thought of as engaging the interaction of both hemispheres. Davidson (1995), for example, hypothesizes that the left and right anterior regions of the brain are key components of an affective regulatory system for approach and avoidance behaviors.

It is with this executive programming we acquire the ability to name, to comment upon, to ***generalize***[36] and to ***choose*** between our contending sets of behavioral impulses: self-preservation, commonly called, at a high level of cognitive generalization, "egoistic" or "self-interested" behavior, and affection, which we call, at an equally high level of cognitive generalization, "empathetic" or "other- interested" behavior. ***Empathy allows us the critical***

[36]The ability to self-consciously generalize is apparently a unique gift of the neocortex with its billions of neurons interconnected into hierarchical networks. The level of generalization issue in all our disciplines likely springs from this. That is, we can move from parts to wholes in generalizing and from wholes to parts in analyzing freely up and down throughout our neural networks. Generalizing (and implicitly, analyzing) has been recognized by scholars in many disciplines as perhaps the defining characteristic of the human brain (e.g., Hofstader 1995: 75; Einstein 1954: 293). This generalizing capacity loosens up the tight wiring of routines and characteristics of earlier brain structures and allows us to manage and, to some degree, overcome the mechanisms that we inherited in common with kindred species (e.g., see Panksepp 1998: 301). In other words, the generalizing, analyzing capacities of the neocortex change the rules of the game for us humans by freeing us up from the blind tyranny of primitive mechanisms. The capacity to generalize is closely related to the issue of cortical plasticity. For a review of current research see Buonomano and Merznich (1998) and Kolb (1995: esp. p. 10). Kendler (1995) sees a layered neural system at a lower level wired tightly for automatisms and a higher layer that supports more flexible, creative processes. This generalizing, higher-level capacity must always be weighed when trying to apply findings in, for example, even primate ethology to humans. One of the reasons our feelings and motives are so difficult to verbalize and communicate to others is probably because the earlier evolved brain (reptilian and limbic) systems are nonverbal. Their input enters the neocortex through neural pathways as inarticulate urgings, feelings. It falls to the neocortex with its verbal and generalizing ability to develop words and concepts to attempt to understand and convey these inarticulate urgings. MacLean (1992: 58) states that the triune brain structure provides us with the inheritance of three mentalities, two of which lack the capacity for verbal communication.

social capacity to enter into or respond emotionally to another's self-interest as well as other emotional states.[37]

In other words, our executive programming, especially our frontal cortex,[38] has the capability and the responsibility for cognitively representing these limbic and protoreptilian brain inputs and making what may be thought of as our ***rational*** and ***moral*** choices among our conflicting, impulsive, and irrational or nonrational motivations. This self conscious, generalizing, choosing capacity accompanied, of course, with language, is what differentiates us from even closely related primate species and makes findings in primate behavior, although highly interesting and unquestionably important, insufficient in themselves to fully understand and account for human behavior.

EXECUTIVE PROGRAMMING, NEURAL NETWORKS, AND NEURAL GLOBAL WORKSPACE

Bernard Baars, of the Wright Institute and his colleagues have proposed a Neural Global Workspace Model (GW) which combines the concepts of

[37]My use of the term empathy here includes the affectional feelings of sympathy which are dependent upon empathy, plus cognitive aspects (Hoffman 1981). Losco has noted that empathy, amplified by cognitive processes, could serve as an evolved mediator of pro-social behavior (1986: 125). Empathy and sympathy are frequently used inclusively, especially in more recent writing (Eisenberg 1994, Batson 1991). For this reason, to include the aspect of sympathy, I have chosen the less common, but analogous spelling of *empathetic* rather than the more usual *empathic*. The positing of the ego and empathy dynamic goes back to the historical juxtaposition of self-interest or egoism and sympathy or fellow feeling in the thought of David Hume, Adam Smith, and Schopenhauer (Wispe 1991). The present articulation goes back to my doctoral dissertation done at Stanford University (1974). The conflict systems neurobehavioral model was applied in several programs which I authored for corporate management training through the education and consulting corporation United States Education Systems during the period 1976-85. Recently, Roger Masters (1989) has also noted the possible innate roots of contradictory impulses to include selfishness and cooperative or altruistic behavior in human nature. De Waal and Aurelli (1997) see empathy emerging in the great apes and as an essential ingredient in conflict resolution that finds continuity with humans. Trudi Miller (1993) has also drawn our attention to this historical duality and suggested its applicability for today. Neither Hume, Smith, Schopenhauer, Wispe, Masters, nor Miller, however, attempted to articulate a model of behavior based upon this duality, or as MacLean calls it, "triality", acknowledging the role of the neocortex in articulating the otherwise nonverbal urgings (1993).

[38]The frontal neocortex especially has long been recognized to be involved in executive functions. See the excellent summary and discussion of findings in Fuster (1997: 150-84). See also Pribram (1973, 1994). Although executive function is frequently equated with frontal cortex function, Eslinger (1996) reminds us that the neural substrate of executive functions is better conceptualized as a neural network which includes the synchronized activity of multiple regions, cortical and subcortical (1996: 392). Eslinger also notes the usual neglect of critically important affectively based empathy as well as social and interpersonal behaviors in neuropsychological, information-processing, and behavioral approaches(390-91).

attention, working memory, and executive function into a theatre metaphor. Baars and colleagues (Newman, et al. 1997; cf. Harth 1997) review other neuroscience and neural network models that deal with attention, binding, resource allocation, and gating that share significant features with their own GW model for conscious attention (for an alternative model based on an evolutionary and clinical approaches and which draws upon MacLean's triune concept, see Mirsky 1996).[39] The authors acknowledge that the models they present implement only partial aspects of their GW theory. Notably neglected are the influences of memory and affective systems upon the stream of consciousness (1997: 1205). Other cognitive metaphors, compatible with GW theory, like Minsky's "Society Theory"(1979) and Gazzaniga's "Social Brain" (1985), remain cognitive in their treatment of sociality although they may be taken to imply affective mechanisms. The CSN model presented in this book attempts to incorporate the affective (generalized into empathy) neural substrate necessary to initiate and maintain sociality.

It is noteworthy that distributed artificial intelligence (DIA) models more closely approximate interpersonal behavior in that they seem to reflect an effort at intelligent balance between the competitive self-interest and cooperation which is necessary to the operation of complex social organizations (Newman, et al. 1997: 1196; Durfee 1993). Underpinning the CSN model, the neural substrate for self-survival (generalized as ego) mechanisms may proceed from circuits in the basal ganglia and brainstem (protoreptilian complex) through connections with the amygdala and other limbic structures (early mammalian complex) which add emotion or passion (see Kandel, et al. 595-612), ultimately to be gated into the frontal cortex by thalamocortical circuitry (e.g., see LaBerge 1995; Crick 1994; Baars 1997, 1988).

Likewise, the mammalian nurturing (affectional) substrate and its associated motivation, a fundamental component underlying empathy, may originate in the septal and medial preoptic limbic (see Fleming, et al. 1996; Numan 1994; Numan and Sheenan 1997) areas, proceed through hippocampal and other limbic structures, and, in turn, be gated into the frontal cortex by neuromodulating thalamocortical circuits (to include the cingulate cortex), where the conflict with egoistic imputs is resolved in the executive or Global Workspace of conscious self-awareness.

The neuromodulating and gating of affect, as well as cognition, by the thalamocortical circuitry is supported by neurologists Devinsky and Luciano (1993) who report that the limbic cingulate cortex, a cortical structure

[39]Levine(1986) has also considered MacLean's triune modular concept as a useful tool in network modeling.

closely associated with the limbic thalamus, can be seen as both an amplifier and a filter, which joins affect and intellect, interconnecting the emotional and cognitive components of the mind (1993: 549). Tucker, Luu, and Pribram (1995) speculate that the network architecture of the frontal lobes reflects dual limbic origins of the frontal cortex. The authors suggest that the two limbic-cortical pathways apply different motivational biases to direct the frontal lobe representation of working memory. They further suggest that the dorsal limbic mechanisms which project through the cingulate gyrus may be influenced by ***social attachments***, and that such projections may initiate a mode of motor control that is holistic and impulsive. On the other hand, they speculate that the ventral limbic pathway from the amygdala to the orbital frontal cortex may implement a tight, restricted mode of motor control reflecting the adaptive constraints of ***self-preservation*** (1995: 233-34). This is consistent with the CSN model in which ego and empathy represent conflicting subcortical inputs into the cortical executive. Several researchers have posited the dynamic of conflicting modules, vying for ascendency in behavior and consciousness (e.g., Tonomi and Edelman 1998; Edelman 1992; Dennett 1998; Pinker 1997).

ELECTROCHEMISTRY, HORMONES AND NEUROTRANSMITTERS

Although it is beyond the scope of this chapter to attempt to deal with the as yet partially understood detailed electrochemical physiology of the egoistic/empathetic conflict, it is appropriate to acknowledge that such behavior is made possible in part by the complex electrochemical excitatory and inhibitory interactions among groups of interconnected neurons (e.g., see the discussions in Cowan, et al. 1997; Fuster 1997: 102-149; Gutnick and Mody 1995). The role of hormones and neurotransmitters must also be acknowledged in any complete analysis. For instance, from the egoistic perspective, testosterone is associated with competitiveness and power urges. Serotonin levels in humans seem related to confidence and self-esteem. On the empathetic side, oxcytocin, arginine vasopressin, and prolactin are important to pair bonding and maternal as well as paternal caring behavior. Opioids (endorphins and enkaphalins) seem important to positive social relationships. For readers interested in more detail, two recent and wide-ringing volumes update the research focusing specifically on affiliation and affection: Carter, et al. (1997), *The Integrative Neurobiology of Affiliation* and Panksepp (1998), *Affective Neuroscience*. Panksepp especially speculates on the contrast between testosterone-driven power urges and oxytocin and opioid mediated affectional behavior (1998: 250-259). Damasio reminds us, however, that there is a popular tendency to overemphasize the efficacy of hormones by themselves. Their action depends upon neural architecture and their effects may vary in different brain regions (1994: 77-78).

THE CONFLICT SYSTEMS NEUROBEHAVIORAL MODEL AND THE THEORY OF MIND

The CSN model presented here may suggest comparison in some respects to the *theory of mind* concept which has become popular in some quarters of cognitive neuroscience in recent years. The term was originally coined by primatologist David Premack (Premack and Woodruff 1978) researching the question of whether chimpanzees had a concept of other minds existing in their fellow primates. It has since been used in an attempt to account for the deficit of relatedness to others presumed to be central to the condition of autism (Frith 1997, 1989; Baron-Cohen 1995; Brothers 1995). It has been applied to child development, where the standard account is concerned with the child's grasp of others' attention, beliefs, and false beliefs (Astington, et al. 1988). Attention has also been directed toward how the child constructs the meaningful intention and evaluative attitudes of others (Fridlund 1991; for a critique of some current issues, see Grossman, et al. 1997).

Any adequate theory of mind would have to have to allow, either explicitly or implicitly, for a generalized concept of self or ego as well as a similarly generalized concept of empathy or other interest...or else it would be utterly meaningless. That is, to have a theory of the mind of others, you must first have an idea of a mind of self. The two exist only in contrast to each other as part and parcel of the fundamental, primary algorithmic subject/object distinction become self-conscious. The theory of mind has been looked upon by some in evolutionary psychology as a specific and discrete evolved brain module (e.g., Cosmides and Tooby 1992). The CNS model presented here, of course, takes a much broader evolutionary view. It sees a sense of self and a sense of otherness proceeding from the inputs of historically evolved and interconnected protoreptilian and mammalian structures generalized by the massive computing and processing power of hierarchically-structured and interconnected neural networks of the spectacularly expanded primate and human neo (or iso-) cortex.

In spite of its vast complexity, the human brain remains essentially the three-dimensional protein structure as well as behaviorally active expression of the genetic informational matrix. The brain/mind/body connection remains inviolate. The primary algorithmic synthesizing loop, multiplied, amplified, variegated, and interconnected, maintains its fundamental matrix-product configuration. Learning, modification of behavior, is essentially modification of the protein expression which constitutes the soma or body. The somatic connection...to include self-preservational and affectional somatic mechanisms as proposed in Damasio's somatic marker hypothesis...

maintains. Of course, as noted previously, in the returning sector of the algorithmic loop from protein to genetic matrix, such individual or phenotypical learning or behavioral modification is not passed on to the DNA matrix because of the barriers inherent in the protein molecular configuration.

Chapter 6

THE ALGORITHMIC RULES OF RECIPROCAL BEHAVIOR

Giving, itself, is one of the strongest sociological functions. Without constant giving and taking within society – out side of exchange, too – society would not come about...Every act of giving is, thus, an interaction between giver and receiver
(G. Simmel, Sociologist, 1950: 389)

The two master, inclusive and modular programs of self-preservation and affection that have been wired into our brain structure operate dynamically according to a set of subjectively experienced and objectively expressed behavioral rules, procedures, or algorithms. Understanding the workings and applications of these algorithms is the key to grasping the role of dialectical conflict and stress in our subjective experience as well as in our interpersonal lives at all social levels.

The major ranges of the conflict systems neurobehavioral behavioral model (figure 6-1) illustrate the features of this ego-empathy dynamic. In the display, subjectively experienced internal as well as interpersonal behavior is divided from right to left into three main ranges called the egoistic range, the dynamic balance range, and the empathetic range. Each range represents a particular mix of egoistically and empathetically motivated behaviors. The solid line stands for ego and pivots on the word "ego" in the executive program of our brain diagram. The broken line stands for empathy and pivots on the word "empathy" in the diagram.[40]

[40]The dynamic of the model, the tug and pull of ego and empathy, self-and other-interest, allows the expression of the mix of motive and behavior as a range or spectrum. The usual dichotomizing of self-interest and altruism is seen only at the extremes of the ranges. All or

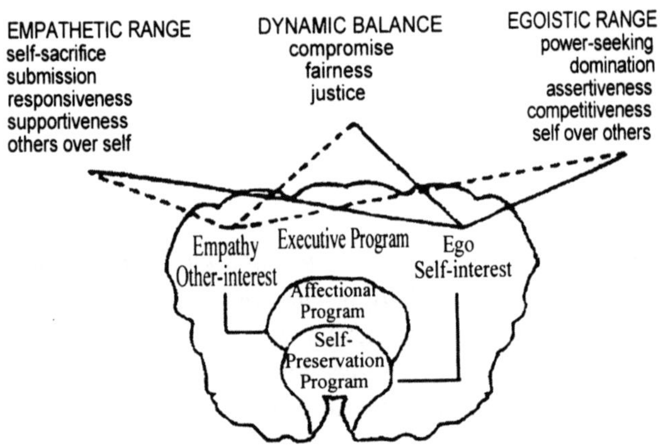

Figure 6-1. The major ranges/modes of behavior.

THE EGOISTIC RANGE

The egoistic range indicates behavior dominated by self-preservation programming. Since the two behavioral programs are locked in inseparable unity, empathy is present here, but to a lesser degree. Behavior in this range is self-centered or self-interested and may tend, for example, to be dominating, power-seeking, or even attacking, where empathy is less. When empathy is increased, ego behavior will become less harsh and may be described more moderately as controlling, competitive, or assertive. As empathy is gradually increased, the intersection of the two lines of the diagram will be drawn toward the range of dynamic balance. Ego behavior

---

most of behavior is a mix of varying proportions. Jencks (1990: 53-54) also notes that every motive or act falls somewhere on a spectrum or range between the extremes of selfishness and unselfishness. Teske (1997) sees a blend of self- and other-interest is his identity construction concept. The CSN model moves to identify and explicate some fundamental brain algorithms that provide framework, structure, and dynamic to our socio-experiential performance. At this point it is perhaps also clarifying to acknowledge that the neural mechanisms underlying social behavior may vary among unrelated species to the extent of being entirely different when we move, for instance, from the relatively simple neurological structures of social insects, who apparently function like automatons, to the enormous complexity of the human brain which functions on the basis of choice among conflicting alternatives.

will be softened as empathy is added. But the defining characteristic of the egoistic, self-interested range is self-over-others. Whether we are blatantly power-seeking or more moderately assertive, in this range we are putting ourselves, our own priorities, objectives, and feelings, ahead of others. We're telling others, "me first."

THE EMPATHETIC RANGE

The empathetic range represents behavior weighted in favor of empathy. Ego is present, but is taking a back seat. When ego is present to a minimal degree, empathetic behavior may tend to extremes of self-sacrifice and submission. When ego is increased, empathetic behaviors become moderated and may be described as supportive, responsive, or any of a variety of "others first" behaviors. As the influence of ego is gradually added, empathetic behavior will approach the range of dynamic balance. In the empathetic range, the key phrase to remember is others-over-self or others first. Whether we are at the extreme of self-sacrifice or more moderately responsive, we are putting the priorities of others ahead of our own.

THE DYNAMIC BALANCE RANGE

The range of dynamic balance represents a working balance between ego and empathy. At this point our behavioral programs are operating in roughly equal measure. I speak of "working," "rough," or "dynamic" balance because the tug-and-pull between the two programs continues ceaselessly. The dynamic nature of the programming means that "perfect" balance may be a theoretical point, unattainable in practice. Our more balanced behavior tends to be characterized by equality, justice, sharing, and other behaviors that show respect for ourselves and others. In fact, respect for self and others is the keynote of the range of dynamic balance.[41] We are working to achieve shared priorities, objectives, and feelings.

ENERGY OR ACTIVITY LEVEL

The extent to which the master modules or programs of self-preservation and affection, ego and empathy, are out of balance, or pulling against each other, is a measure of behavioral tension. We experience this behavioral tension both internally and between ourselves and others, in any relationship or interaction. Unmanaged or excessive tension becomes, of course, behavioral stress. But that's not all. Important also is the degree of energy we give to the interaction or the relationship. The amount of energy we put into any activity depends mostly upon how important we think it is or how

[41]See Eckel and Grossman (1997). Without making any connection with brain science or the reciprocal rules of behavior, the authors use a typology of fairness (for me, for you, for us) which expresses the conflict systems model and the reciprocal algorithms of behavior.

enthusiastic we feel about it. In competitive sports or contests, qualitative differences in energy are easily observed. In intellectual contests, like chess, the energy may be intense, but less obvious. Although self-preservation and affectional programs--ego and empathy--are the main variables of interpersonal behavior, energy or activity level is often important to consider in describing more fully our interactions.

THE PROPOSED OPERATING ALGORITHMS OF INTERPERSONAL BEHAVIOR

From the dynamic interplay of ego, empathy, and activity level come the following rule statements.

1) Self-interested, egoistic behavior, because it lacks empathy to some degree, creates tension within ourselves and between ourselves and others. The tension increases from low to high activity levels. And it increases as we move toward the extremes of ego.

Within ourselves, the tension created by the tug of neglected empathy is experienced as a feeling of obligation to others or an expectation that they might wish to "even the score" with us.

Within others, the tension created by our self-interested behavior is experienced as a feeling of imposition or hurt, accompanied by an urge to "even the score."

Children often reveal the dynamic of such behavior in a clear, unsophisticated form. Imagine two children playing on the living-room floor. One hits the other. The second child hits back, responding in kind. Or the children may not hit each other at all. One might instead call the other a bad name. The second child reciprocates, kicking off a round of escalating name-calling. One child may eventually feel unable to even the score and will complain to a parent to intervene. Most of us have experienced such give-and-take as children and have seen it countless times in our own children and grandchildren. Similar behavior is embarrassingly observable among adults. It can be seen in husband-and-wife arguments, bar fights, hockey games, political campaigns, even in sophisticated lawsuits. The rule operates not only in such highly visible conflict situations, but also in very subtle interactions--in the small behavioral exchanges, the ongoing give-and-take of all interpersonal relations.

Expressive of the underlying conflictual excitatory/inhibitory dynamic of the neural architecture, we can say that

the reactions that build in ourselves and others do so potentially in proportion to the behavioral tension created by the egoistic, self-interested behavior.

That is, the harder I hit you, the harder you hit me in return. Or the fouler name you call me, the fouler name I call you in return. Or perhaps with more sophistication, I resolve the tension in me by an act of visible "superiority." I ignore you--although I could call you an even fouler name, if I chose.

Behavior on the other side of the scale is described in the second rule statement:

2) ***Empathetic behavior, because it denies ego or self-interest to some degree, also creates tension within ourselves and others. This tension, likewise, increases as activity levels increase and as we move toward the extremes of empathy.***

Within ourselves, the tension created by the tug of neglected self-interest (ego) is experienced as a feeling that "others owe us one" and a growing need to "collect our due." This tension, especially if it continues over time, may be experienced as resentment at being exploited, taken for granted, not appreciated, or victimized by others.

Within others, the tension created is experienced as a sense of obligation toward us.

The reactions that build in ourselves and others, again, are in proportion to the behavioral tension created. And again, the unmanaged, or excessive tension is experienced as behavioral stress.

When we do things for others--give them things, support them, make personal sacrifices for them--it can make us feel good, righteous, affectionate, loving. But when we do these things, we do want a payback. That's the tug of self-interest. It can be very slight, hardly noticeable at first. But let the giving, the self-sacrifice, go on for a while, unacknowledged or unappreciated (that is, without payback to ego), and see how we begin to feel. The tension, the stress, starts to show. We complain that others are taking advantage of us, failing to appreciate us, taking us for granted, victimizing us. Self-interest cannot be long short-changed without demanding its due. We may eventually relieve the stress by blowing up at those we have been serving--accusing them of ingratitude, withdrawing our favors, or kicking them out of the house. Or we may sandbag the stress, letting it eat away at our dispositions, our bodies.

On the other hand, when we do things for others, they often feel obliged to return the favor in some form to avoid being left with an uneasy sense of debt. Gift-giving notoriously stimulates the receiver to feel the need to reciprocate. Think of the times when you have received a holiday gift from someone for whom you had failed to buy a gift. Sometimes the sense of obligation prompted by the empathetic acts of others can become a nuisance.

The third rule statement describes the balance between these two extremes:

3) ***Behavior in the range of dynamic balance represents an approximate balance of ego and empathy. Within ourselves and others, it creates feelings of mutuality and shared respect.***

For most of us it's an especially satisfying experience to interact with others in equality, with no sense of obligation, superiority or inferiority. To work together in common humanity, in common cause, is to experience behavioral dynamic balance. Of course, there are many versions of the experience of dynamic balance: the shared pride of parents in helping their child achieve, the joy of athletes in playing well as a team, the satisfaction of co-workers in working together successfully on an important project.

THE RECIPROCAL NATURE OF BEHAVIOR

These algorithms of behavior operate in the smallest interactions, the vignettes, of everyday personal life. The dynamic of behavioral tension dictates that for every interpersonal act, there is a balancing reciprocal. A self-interested act requires an empathetic reciprocal for balance. An empathetic act, likewise, requires a balancing self-interested reciprocal. This reciprocity goes back and forth many times even in a short conversation. Without the reciprocal, tension builds, stress accumulates, and either confrontation or withdrawal results.

RECIPROCITY THROUGH CONFLICT.

These, then, are the proposed basic interpersonal algorithms of our three-level brain. These algorithms show how we get to reciprocity through conflict. I propose that they shape the conflict and reciprocity, the give- and-take, at all levels of our interactive, social lives.

Overemphasis on either self-interest or empathy, exercise of one program to the exclusion of the other, creates tension and stress in any social configuration -- from simple dyadic person-to-person encounters up to and including interactions among members of the workplace, society at large, social groups, and entire economic and political systems.[42]

[42]Somit and Peterson (1997) see that evolution has provided us with a predisposition for hierarchically structured social and political systems, in other words a tendency to hierarchy. I would suggest that this may be seen as an alternative perspective of the same dynamic of the tug and pull of ego and empathy of the triune modular brain structure that underlies reciprocity. Reciprocity, although more often than not seemingly unbalanced, in social and political relationships, is nevertheless always there to some degree. Even the range of dynamic balance of the conflict systems model is an approximate and shifting balance with some degree of hierarchy of dominance and submission. In its purest form, ceteris paribus, the innate dynamic only tends, rather imperfectly toward a balanced reciprocity. It does not and cannot achieve it deterministically, but only probabilistically. When other things are not equal, i.e. there are differences in personal strength, talent, ability, and intelligence, such differences will allow some individuals or groups to control more resources and thereby create hierarchies. Such hierarchies may be accepted by the less capable, but not without behavioral tension. The hierarchies are inherently unstable because of this behavioral tension. That is, as soon as the unequal capacities become less unequal, those on the lower end almost invariably move to contest and alter the hierarchy. Somit and Peterson acknowledge this tendency to unbalanced reciprocity or hierarchy as regrettable. And they devote their book to

THE QUESTION OF SCIENCE: PHYSICS VS. SOCIAL

The proposed reciprocal algorithms of behavior can be viewed as high-level brain algorithms, built up from a nested hierarchy of interconnected lower-level algorithmic modules.[43] The algorithmic rules, as proposed here, operate very imperfectly. I suspect that this will be true of any behavioral algorithms or principles, proposed at this level of generalization. The proposed algorithms, then, can approximate, but not fully achieve, the precision of the laws of classical physics or even quantum mechanics. This is because they are achieved through the process of organic evolution (which involves some chaotic and random processes as well as natural selection) and therefore cannot operate as immutable universal physical laws but as generalized algorithms with degrees of variation.

The idealized, or rather statistically generalized, tug and pull of ego and empathy presented here may be further probabilized in actuality by genetic, gender and developmental, individual experience and learning, and other environmental shaping and reinforcing factors.

In other words, genetically speaking, given the individual differences in genetic inheritance that we see in such obvious things as in hair, skin, or eye color, some individuals behaviorally may be more or less as strongly wired for self-preservation and affection as others. And granting gender and developmental differences, every human being is, nevertheless, similarly wired with the fundamental brain architecture unless he/she has very serious genetic defects indeed. Influential developmental psychologists like Jean Piaget (1965) of Switzerland and Lawrence Kohlberg (1984) of Harvard, operating from a behavioral perspective, have constructed and tested theories of childhood moral development. In the theories of both men moral stages

---

helping us understand how we may achieve the desired, balanced or non-hierarchical political system. Salter (1995) has pulled together a considerable quantity of naturalistic observation on the genetically-based communicative signaling in human and nonhuman species involved in the negotiating and maintenance of hierarchies.

[43]The algorithms of reciprocal behavior may also be thought of as high level Darwinian algorithms (see Cosmides and Tooby 1989 and Tooby and Cosmides 1989), which function as the cognitively generalized function (not necessarily the sum) of perhaps many contributing and perhaps more highly specific innate algorithms (see also, Cory 1996; cf. Vandervert 1997). It might be further noted that the CSN model, which rests upon evolved algorithms of the brain, may be consistent with the sensory motor approach to cognition (e.g., see Newton 1996) as long as the very extensive and complex algorithmic processing and structuring of sensory and motor inputs and outputs is not treated too simplistically. For example, Harth comments that the massive loops of reciprocal connectivity between the cortex and the subcortical relays in the visual system give the impression that "the cortex is more bent on introspection and confabulation than forming an unbiased view of the outside world"(1997: 1245).

of development emerge much the same in all cultures when the child experiences anything approaching a normal family life. Such generalized moral stages could not be found across cultures if they were not genetically based on the species-wide brain structure and its associated behavioral potentialities.

From the standpoint of individual learning, socialization, and other environmental factors, modifications in biological structures and potentialities occur in early development and throughout life. Individual life experiences may facilitate, suppress, strengthen, or otherwise channel, the expression of these inherited biological programs. Environmental factors, to include physical constraints as well as our socially and scientifically accepted institutions and paradigms, may also shape and reinforce the expression of the evolved algorithmic dynamic. Individual learning, experience, or environmental factors of the individual life cannot, however, eliminate the genetic structure and programming of the brain...that is, not without radical injury, surgical or genetic intervention. And the behavioral tension will be there to both resist the changes and to shape the experience, even shape the environment itself, in a dynamic manner.

Because of these factors, the behavioral algorithms are statistical...much in the same way as are the second law of thermodynamics and quantum theory of physics. That is, they do not allow precise prediction of specific behavior at the basic unit of analysis...the individual, molecular, or subatomic level respectively...but only on the aggregated basis of statistical probability. The proposed algorithmic rules of reciprocal behavior, as here presented, may nevertheless very well prove to be equally as valid and useful to social science as the laws of physics are to physical science. They do not and cannot, however, have the immutable quality of physical laws such as gravity. As products of organic evolution, they inevitably involve more probabilities because of individual differences, genetic and learned, in the evolved basic units.

It is also interesting to note that an analogy can be made between the inclusive spectrum of possible behaviors of the conflict systems neurobehavioral model and the particle/wave function of quantum physics. As the wave function of a particle is defined to include all the possible values of a particle according to probability, the wave function of behavior can be thought to include all possible internal and interpersonal behavioral probabilities (mixes of ego and empathy) extending across the egoistic, empathetic, and dynamic balance ranges. Externally, observed behavior is predictable from the model as is quantum behavior, only on a probability basis specified by the wave function. The behavioral wave function, like that of particle physics, collapses or reduces to one behavior in a decision, action, or observation. If it doesn't collapse, we see frustration, tension, and indecisiveness... ambiguous behavior stalled in uncollapsed waveform.

Upon observation by an external observer, the wave function of behavior can be considered to collapse to a specifically observable behavior on the part of the individual and that's the end of it. But internally, we experience it differently because we have access to the dynamic. We know, in our conscious awareness, the tension, the difficulty, the struggle we go through in important issues of ego and empathy conflict. In physics, however, we, perhaps, simply do not yet understand what set of dynamics leads to the wave function collapse.[44] In behavior, the dynamic lies in the complexities of subjective preconsciousness and/or self-aware consciousness.

THE CONFLICT SYSTEMS NEUROBEHAVIORAL MODEL VS THE MASLOW HIERARCHY

It is useful at this point to return to the Maslow need hierarchy and contrast it with the conflict systems neurobehavioral model developed by building upon the triune brain model of MacLean.

Maslow's focus on a staircase-like hierarchy of inner needs tended to turn us inwardly away from the social environment. Co-opted and blended with the prevailing view of self-interest as the dominant human motive, Maslow's theory of self-actualization, with its lofty connotations of self-fulfillment and creative expression became reduced, especially in the decades of the 70s and 80s, to a license for indulgent self-interest (Yankelovich 1981; Cory 1992). This isolated, indulgent version of self-interest, as expressed in our social and business experience, earned the labels of "narcissistic" and "me first." One of the great popular appeals of the Maslow hierarchy is that it fit so well with the prevailing emphasis on self-interest in our everyday as well as academic thinking on economics and politics.

CONFLICT, NOT EMERGENCE.

But inward focus and simplistic hierarchy were not the only problems with Maslow's hierarchy. It allowed us to be drawn excessively toward our

[44]That is, in physics it is not known exactly why and how wave function collapses or reduction occurs and how eigenstates are determined (e.g., see Hameroff and Penrose 1996: 311). The standard Copenhagen Interpretation sees collapse as occurring at randomly measured values when the quantum system interacted with its environment, was otherwise measured, or consciously observed; (e.g., see Stapp's 1972 well-known article on the Copenhagen Interpretation). Penrose (1994) and Hameroff and Penrose (1996) introduce a new physical ingredient they call objective reduction (OR), which becomes guided and tuned into orchestrated OR, in which quantum systems can self-collapse by reaching a threshold related to quantum gravity. Ellis has compared consciousness to a wave pattern or function (1986: 67). Harth notes, in summarizing his sketchpad model, that "the transformation from the extended activities in the association areas and working memory to specific mental images may be likened to the collapse of a wave function in quantum mechanics..." He does not, however, imply any quantum effect(1997: 1250).

inner selves and away from society because it missed a point central to human behavior--that conflict, not emergence, is behavior's most definitive characteristic.

Maslow placed the social or relatedness needs (empathy) lower on the escalator than esteem and self-actualization needs (ego). His theory contains the clear suggestion that we pass through these social needs or rise above them in the trek up the hierarchical ladder. This was a fundamental error, resulting in a considerable distortion of our view of human social nature. The two sets of needs, social and self-interested, although hierarchical to some extent, are wired together in the same brain to produce the tug and pull, conflictual dynamic of interpersonal or social behavior. We are not autonomous, but at best semi-autonomous creatures, completely immersed in a pervasive social context, which is, at this point in our evolution, both demanded and made possible by our evolved brain structure.

FOR ME? FOR YOU? OR FOR US?
IT'S ALWAYS THE SAME!

The tug and pull of ego and empathy drives interpersonal behavior not only at the levels of Maslow's social and esteem needs, but at all levels. The basic questions of interaction are the same anywhere in the hierarchy.

These questions are: Do I do it or take it for myself (ego)? Do I do it for or give it to others (empathy)? Or do I do it for both myself and others? (dynamic balance).

It is enlightening to try out these questions at each level of Maslow's staircase. Take the physiological needs first. I can see myself lost in the desert with a friend. I have one remaining canteen of water. It's half full, and I don't know if or when we'll find more. My interpersonal choices are three: Do I keep it all for myself (ego)? Do I give it all to my friend (empathy)? Or do we share it (dynamic balance)? I would face the same choices if my companion were a spouse, child, friend, or enemy.

These three questions would likewise apply to the next level of needs, those of safety. Suppose we are threatened by a wild beast? Or a natural disaster? An intruder? A terrorist? Do I protect myself? Others? Self and others? The conflict is more or less evident depending on the urgency of what is happening, but it is always present.

At the next level the social needs are directly related to affection and empathy, while above them, the esteem needs are related to ego. But as we have seen ego and empathy and their inexorable conflict pervade all three levels of Maslow's hierarchy. In evolutionary terms, the conflict did not exist prior to the appearance of affectional programming in the mammalian brain; but ever since it appeared, it has influenced all needs and all behaviors.

The ego-empathy conflict even pervades Maslow's highest need level, self- actualization. Behaviorally speaking, the need for self-actualization has to do with ego rather than empathy, but again the choices are the same as at all the other need levels. Do I put my own priorities, feelings, and objectives first? Or do I first consider the priorities, feelings, and objectives of my parents, spouse, children, friends, company, church, nation, or world-wide humanity itself? Or do I struggle to achieve a balance?

CAUTIONS

The two algorithmic processes, the primary and the secondary set, should not be confused with reifications of abstract concepts or vital forces shaping and moving through evolution and/or everyday behavior, like the ***entelechy*** postulated in the 1890s by German scientist, Han Driesch (1929), or the ***elan vital*** of the French philosopher, Henri Bergson (1944). The two algorithms are mechanistically derived: The first, from the protein-nucleic acid synthesizing loop with its related metabolic processes, later replicated in our neurological structure; the second, from our evolved three level modular or triune brain structure.

For purposes of brevity and simplicity, in the chapters that follow, I will refer to the algorithmic rules of reciprocal behavior inclusively as the second or reciprocal brain algorithm. The shaping effect and pervasive influence of the primary and reciprocal brain algorithms will be traced in the chapters that follow across the spectrum of our thought, behavior, subjective experience, and language.

PART TWO

BRAIN ALGORITHMS IN PSYCHODYNAMICS, COGNITION, PHILOSOPHY, AND MYSTICISM

Chapter 7

HOW WE THINK: BRAIN ALGORITHMS IN PSYCHODYNAMICS

Only intelligence... by action, and by thought, tends toward an all-embracing equilibrium by aiming as the assimilation of the whole of reality
(Piaget, Psychologist, 1947: 9)

The primary and reciprocal algorithms describe not only how we fundamentally behave, but also how we think. The primary algorithmic pattern of ***order →chaos →order*** and its variant, ***incorporation → extension → incorporation*** appear throughout the discipline of psychology mostly unintentionally and in a disguised form. The reciprocal ego/empathy dynamic, likewise, is expressed, sometimes explicitly sometimes not. The two algorithms often simply pop out, in part and here and there, in speaking about how the mind works.

FRIEDRICH NIETZSCHE: THE WILL-TO-POWER AS EXTENSION AND INCORPORATION

Of the 19th Century philosophers who most influenced modern psychology, Friedrich Nietzsche (1844-1900) intuited the primary algorithm of extension and incorporation to be the pervasive pattern of life. His descriptions of the will-to-power, which he mistakenly saw, not as an algorithm, but as a vital force driving all life, show these features clearly. For example, Nietzsche states that the will-to-power is expressed in

> ...the primeval tendency of the protoplasm when it ***extends*** pseudopodia and feels about. ***Appropriation*** and ***assimilation*** are above all a desire to overwhelm... (emphases in the passage are mine) (Nietzsche 1968: 346, see note 656, Spring-Fall 1887).

Nietzsche was apparently describing the incorporative act of the one-celled microorganism, the ameba, as it encircled and engulfed a piece of matter. Instead of understanding the act as a behavioral algorithm of organic synthesis, as I previously pointed out (ch. 4: 29), he assigned it, in keeping with the vitalistic thought of his day, a cosmic or vital source which he called the will-to-power, a force that seeks to incorporate everything.

Further on, he writes:

> What is "active"?–***reaching*** (emphasis mine) out for power.
> "Nourishment"--is only derivative, the original phenomenon is the desire to ***incorporate*** (emphasis mine) everything"(Nietzsche 1968: 347: see note 657, Spring-Fall 1887)

That Nietzsche is perceiving the primary algorithm of ***incorporation→ extension→ incorporation*** is clearly indicated in the above quote. His will-to-power continuously extends and incorporates, driven by a desire to incorporate everything. He also connects the pattern with eating, but reverses the causality when he sees nourishment as only derivative of the vital force.

The will-to-power was the pervasive motive and shaping theme that underpinned Nietzsche's final philosophy.[45]

INCORPORATION IN THE THOUGHT OF SIGMUND FREUD

The founder of psychoanalysis, Sigmund Freud, used the term incorporation extensively and implied both extensionality and incorporation even when he didn't mention them specifically. A good example is Freud's first period of psychosexual development, the oral incorporative. Freud compares the incorporating of knowledge or other intangibles to taking things in through the mouth. He speaks of a hunger for knowledge and such things as though they were foods that could be ingested (cf. the statement quoted from biologist Margulis in ch. 4: 31).

Freud, then, suggests a clear analogy or connection between the physical metabolic pattern of incorporation-extension and cognition when he explains that the incorporative attitude which comes out of a child's gaining pleasure from taking things into the mouth may extend to incorporation of symbolic things, such as love, knowledge, money, material things, etc.

Freud's influential three-part personality schema of id, ego, and superego shows clearly the metabolic features I have been talking about. The id, the original source of all psychic energy, is the locus of biological drives or instincts. The ego, which Freud thought of as the executive agent of the

[45]Nietzsche's earlier philosophy was dualistic, i.e. the tug and pull between Dionysus and Apollo, wastefulness and purpose, chaos and order. This dynamic was, of course, another expression of the primary algorithm. The unifying theme of the will to power was first published in *Zarathustra*. See Kaufmann (1968: 178-207) on Nietzsche's discovery of the will-to-power.

personality, draws psychic energy from the id through the mechanism of identification. Identification has been defined as the process of incorporating the qualities of an external object, usually a person, into one's personality. Identification, which figures prominently in Freud's system, is therefore incorporative in nature and includes an implied extension into the environment (Hall 1954: 74-78; also Freud 1959: 37-42). Incorporation, in fact, runs all through Freud's writing. A fundamental of his theory, the object-cathexis principle, is the investing, or canalizing of psychic energy toward an object, is an incorporative device.

Although there is plenty of explicit as well as implicit incorporation and implied extensionality in Freud's writing, he never unifies these two functions into the primary algorithm of ***incorporation→extension→ incorporation.*** It is easy for us, however, to make the specific connection from his writing.

Probably because of the sexual repression that was prevalent in his day, Freud got hung up on the mechanism of sex. His theory slipped into fuzziness when he overemphasized sex and failed to distinguish properly between sexual mechanism and the algorithmic quality of behavior.

ERICH FROMM AND ASSIMILATION

Erich Fromm, a noted psychoanalyst, who followed Freud, gives us his basic ideas about human nature in a volume called, *Man for Himself.* Fromm moved away from Freud's emphasis on sex and stated a more clearly metabolic perspective although he did not explicitly connect up the major behavioral pattern. Fromm defined character, as he saw it in people, as the relatively permanent manner in which human energy is channeled in, what he referred to as, the processes of *assimilation* and *socialization*. According to Fromm, man relates himself to things by acquiring and assimilating them. Acquiring and assimilating are, of course, synonymous with incorporation. This is clearly expressive of the primary algorithm.

As for people, Fromm tells us that it is necessary for man to be related to others, one with them, part of a group (Fromm 1947: 55-67). This is clearly expressive of the reciprocal algorithm of our triune modular brain structure. Fromm, however, doesn't state much of a biological basis for his personality orientations. He just establishes them by a series of assertions which he expects us to take on faith.

Fromm confuses the relation between the physical metabolic act of eating and the cognitive algorithm with sexual overtones is his definitions of sadism and masochism. He sees sadism as originating in the impulse to "swallow" persons, to have complete domination over them. Masochism, on the other hand, is based in the desire to be "swallowed"(Fromm 1947: 113-114).

Fromm's writing never produced a clear, systematic state of either of the two bioneurological or brain algorithms. Intuitively, or rather, implicitly, elements of both underpin his thinking.

FREDERICH PERLS AND DENTAL OR EATING AGGRESSION

Frederich Perls, noted Gestalt therapist, is the psychotherapist who saw most clearly the behavioral significance of the primary algorithm. In contrast to Freud's sexual emphasis, Perls built his theory around the hunger instinct as the guide to behavior. His theory was not well developed and far-fetched at points.[46] Because his perspective was limited to therapy, he did not follow through on the theoretical ramifications.

In *Ego, Hunger, and Aggression*, his main statement of theory, Perls puts forth stages of the hunger instinct much like Freud did with the sexual instinct. For example, concerning the pathology of the impatient eater, he writes:

> ...Above all, the destructive tendency, which should have its natural biological outlet in the use of the teeth, remains ungratified....The destructive function, although in itself not an instinct but a very powerful instrument of the hunger instinct is "sublimated"–turned away from the object "solid food"(Perls 1947: 110).

Perls goes on to say that this sublimated hunger instinct of the impatient eater then manifests itself pathologically in killing, war, and other acts of cruelty, and even may be turned upon the impatient eater himself in torture and self-destruction. Aggression, according to Perls, is mainly a function of the all important hunger instinct with the teeth as the most important biological representation of aggression. He sees dental underdevelopment as a personality defect and warns that such people may remain "sucklings" for life, which is apparently a serious disorder in Perl's scheme of things.

Perls further shows his confusion when he calls for a psychoanalysis of the hunger instinct and mental metabolism. He saw proper *assimilation* at both levels as the key to good health. Assimilation is, of course, a synonym for our preferred term of incorporation. Perls claimed that lack of proper chewing and biting led to biological dental aggression being directed toward

[46]In fact Perls captions the second part of *Ego, Hunger, and Aggression* as *Mental Metabolism.* He apparently was influenced by Smuts, South African general and prime minister, who wrote a theoretical statement arguing that the driving force behind the universe was the creation of greater and more perfect *wholes*. Smuts saw metabolism as an expression of this universal tendency and therefore was vitalistic in his thinking. My position would be that Smuts was contemplating his own bioneurological processes and projecting them out into the universe as a vital force (see Smuts 1926). Perls includes a passage from Smuts at the beginning of Part Two of *Ego, Hunger, and Aggression* (1947: 105).

other people. In addition, it had a direct effect on mental life. At one point, he writes that

> only those who grind their mental food so thoroughly, that they get the full value of it, will be able to assimilate and reap the benefit of a difficult idea or situation(see Perls1947: 115, 117, 122-127).

Perls' theory becomes farfetched because he needlessly forces the connection between mental incorporation and physical incorporation in very literal terms. The act of eating and the act of cognition are ***not*** the same thing, but they do operate according to the same organically-based algorithm. And therein lies the confusion. What Perls doesn't get clear is that each level, physical and mental, functions metabolically in its own right in the ordering or incorporating pattern of the primary algorithm.

EQUILIBRATION AND THE COGNITIVE PSYCHOLOGY OF JEAN PIAGET

Swiss psychologist, Jean Piaget, has been one of the most influential figures in cognitive and educational psychology in the middle to late 20th century. He writes about the similarities and differences between assimilation (incorporation) at the physiological level and at the cognitive level. Unlike Perls, he occasionally confounds the two levels.

For example in his early work, *Psychology of Intelligence*, first published in 1947, Piaget presents us a fundamental definition of process as he sees it when he writes that psychological life begins at the point of functional interaction when:

> ...assimilation no longer alters assimilated objects in a physico-chemical manner but simply incorporates them in its own forms of activity...

Further on in the same paragraph, however, he achieves greater clarity while focusing only on cognition. He writes that subject/object interaction occurs at "ever-increasing spatio-temporal distances" as it develops along increasingly complex paths. In fact he tells us that the entire development of mental activity, from perception, habit, symbolic behavior and memory, up to the higher processes of reasoning and formal thought, is a function of "gradually increasing distance of interaction,"moving toward achieving an equilibrium between an "assimilation of realities further and further removed from the action itself." (Piaget 1968: 7-9).

In addition to incorporation (assimilation), Piaget clearly includes the function of extension in such phrases as ***at ever-increasing spatio-temporal distances*** and in describing the operations of reasoning as taking place at ***gradually increasing distance of interaction.*** In this earlier work, Piaget uses the term ***equilibrium*** to mean a steady or homeostatic state which can include changes, rather than in the sense of a thermodynamic equilibrium which is a state of randomness or maximum entropy.[47]

A final quote from this earlier work gives Piaget's feeling of the ultimate direction of mental activity. He tells us that only intelligence with its capacity for detours and reversals in both action and thought:

> ...tends towards all-embracing equilibrium by aiming at the assimilation of the whole of reality....(1947: 9)

These excerpts from Piaget give us an almost perfect description of what I have called the primary algorithm as it applies to cognitive functioning. The elements of the algorithm are clearly seen, although they are not identified as such. And they can be easily arranged into the process pattern of

INCORPORATION→ EXTENSION→ INCORPORATION→

In his more recent work, *The Development of Thought*, which was first published in 1975, Piaget moves from the use of the term ***equilibrium*** to ***equilibration***.

Equilibration allows better for the concept of movement and growth than the former term. Equilibrations do not return to the same place, but build upon and include what has been previously assimilated. It is a sort of progressive movement to more complex levels of order. Piaget, however, is somewhat hardpressed to explain the basic driving forces of these equilibrations, finding them in external and internal imbalances coupled with a requirement for coherence. He states specifically that in our present state of knowledge it seems difficult to support the proposition that nonbalance or contradictions are inherent in the very characteristics of thought.[48]

This last, of course...the structuring of parts, the resolution of contradictions or imbalances...is in the very nature of the organic synthesizing algorithm. We see the rhythmic dialectical algorithm of ***incorporation→ extension→incorporation*** as generating successive imbalances and involving us in cognitive interaction with the environment where we may encounter further imbalances in a mutually reinforcing interaction.

As active, energized life forms, which evolved in a reciprocal relationship with an environment which required that we make discriminations (of nutrients, and later, threats), we are seekers, explorers of our environment--to seek implies nonbalance or contradiction. This is the way a living organism is structured and, above the most simplistic levels cannot survive

[47]Piaget introduces the additional concept of accommodation which indicates the organism's adjustment to the environmental stimuli, but I prefer to view accommodation as a subfunction of incorporation.

[48]See Piaget (1977: 4-38) for a definition and discussion of the concept of equilibration. On the question of the causes of nonbalance or contradictions, see (12-18).

in any other way. This perception seems preferable to Piaget's position which really can't explain why nonbalances occur--only that when they do, they act as driving forces. In short Piaget observed and described the process with accuracy, he simply lacked the theoretical concepts to explain it adequately. And it troubled him.

In my view, then, as a result of its evolved structure (the evolved subject/object relationship with the environment), the energy and directionality of the primary algorithmic process may well force the discriminations, the contradictions, even when they may or may not be substantially present in the environment. This is consistent with the brain's acknowledged tendency to impose order as noted by Gazzaniga and others (see chapter 3: 24).

If we take Piaget's position all together, we can come up with a dialectical pattern that is a very close match with the primary algorithm. It would be expressed as follows:

EQUILIBRATION→IMBALANCE→EQUILIBRATION

For Piaget, then, intelligence grows and thought develops in a progressive, directional process of equilibration to imbalance to a greater equilibration which includes within itself the resolution of the prior imbalance. Although Piaget does not connect this process directly to the process of organic metabolism and is uncertain of the sources of nonbalances or contradictions, it can be easily seen as a variant phrasing of the same phenomenon...the primary algorithmic process.

PIAGET AND THE SECOND, RECIPROCAL ALGORITHM

Piaget has less success, however, in grasping the reciprocal algorithm, the tug and pull between ego and empathy driven by our evolved brain structure. In his extensive studies of the moral development of children, he deals with the importance of cooperation, mutual respect, and the norm of reciprocity. He feels strongly that his research shows an inherent or biological basis for these behaviors. He contrasts his position to that of the eminent French sociologist, Emile Durkheim, who saw external constraint as the sole source of moral development in children.

And Piaget notes the parallelism between the development of intelligence and morality...between intellectual assimilation and moral assimilation see Piaget 1965: esp., 395-406). Although his moral assimilation process implies, in part, an internal tug and pull between ego and empathy, he never, however, succeeds in separating conceptually the two algorithms.

The first and second algorithms remain confounded in the thought of Piaget.

PHYSICAL METABOLISM AND COGNITIVE OPERATION: THE DIFFERENCES

This section has laid the groundwork for differentiating the functioning of the synthetic algorithm and its related metabolism at the basic organic level and at the cognitive level after the appearance of the highly specialized neural network. Each level functions algorithmically in its own right.

At the basic physicochemical level, the organic algorithm is rather tightly structured and controlled under the guidance of the DNA matrix. A living organism, whether algae or human, can only incorporate so much of its environment. The amount and type of environmental energy and materials it can take in and order is limited by the organism's own evolved essentially protein-nucleic acid structure, but within those limits, established by its genetic code, the organism extends and incorporates in a regular pattern governed by its internal mechanisms. In short, at the physical level, we humans can only eat (incorporate) so much food...some more than others...but we all have clear limits to our capacity to metabolize.

At the cognitive level, there is a significant change. Although we continue to function in an analogous fashion, incorporating new experience and new knowledge, the highly specialized processing capacities of the evolved neural network of the human brain, which magnify and focus the signaling, amplifying, and binding properties of elementary proteins, have greatly extended the limits of the primary algorithmic pattern. As Piaget has noted, only intelligence tends toward an all-embracing equilibrium by aiming at the assimilation of the whole of reality.

Said another way, physically we can devour only a limited amount of the matter and energy around us, whereas mentally we can aim at devouring, assimilating, incorporating all of the universe that our perceptions (extended even by instrumentation), will permit us to experience. This phenomenon is undoubtedly based on the protein structure and action of our highly evolved brain with its billions of neural connections. That is, the mind/body connection is never lost.

Despite the illusion of reaching out and incorporating the whole of existence, cognition or thought remains irrevocably tied to the three dimensional protein structure of our brain and body...and, through this protein structure, ultimately, to the genetic matrix. All thought, subjective experience, and behavior, then, is somatically anchored, ultimately to our survival requirements as set out by the genetic informational matrix. Even in the most abstract of thought processes, the physical matrix is there...the matrix-to-product-to-matrix feature of the primary organic algorithm obtains. Even in thought, something is never created out of nothing.

We might remind ourselves at this point that the primary algorithm of physical and cognitive incorporation, and its varied expressions yet to be identified, may be viewed from an analogical perspective. Whenever we speak of function, or speak functionally at different levels, this is unavoidable. This analogical quality, of course, is also true of our fundamental laws of physics--the laws of thermodynamics. Melting of ice cubes, equalizing of temperatures, entropic movement of gas molecules all have analogous qualities but are still said to express the second law of thermodynamics.

It is important to make the distinction that things can be analogous without having any common grounding or source. The physical and cognitive expressions of the primary algorithm, however, are firmly grounded in the fundamental process of the protein-nucleic acid synthesizing loop with its related metabolic activities as replicated in brain structure. So the patterns can equally be argued as being not only analogous but also derivative.

Chapter 8

HOW WE PHILOSOPHIZE: THE PRIMARY ALGORITHM, DIALECTIC, AND LOGIC IN THE WEST

Unity by itself is not a principle;
in being, unity is multiplicity, but at the cost of self-contradiction.
So much, then, on Plato and quantum theory
(C. F. von Weizsacher, Physicist, 1980: 400))

ORDER→ CHAOS→ORDER

INCORPORATION→ EXTENSION→INCORPORATION

EQUILIBRATION
EQUILIBRATION→ ***→EQUILIBRATION***
IMBALANCE

These are variants of the dialectical algorithm of the organic synthesizing loop expressed from the differing perspectives of physics, biology, psychoanalysis, and cognitive psychology. Although we may conceal it by using synonymous, but different terms as we shift our viewpoint, the algorithmic identity of the patterns is unmistakable. A study of philosophy, east and west, will further bear this out.

Whenever the algorithm is seen as a moving, or process pattern, it inevitably comes up the same dialectical format, explicitly or implicitly. Standing stark and clearly, or obscured by verbiage. When it is viewed in freeze-frame, or still picture, the structural components of the algorithm can be taken apart and examined. From this viewpoint we have only the primary choices of taking things apart or putting them together. In philosophy, taking apart and putting together form the two primary philosophical methodologies. They are called analytics and synthetics.

When we take things apart, we are breaking down a "whole" or higher "order" thereby producing relative chaos, disorder, or imbalance. When we break an idea or a concept down into its parts, we call this ***deductive*** reasoning.

On the other hand, when we put things together, create a whole from a series of separate observations, we are creating order out of the previous disorder or chaos. When we create a concept or generalization from separate facts or observations, we call this ***inductive*** reasoning. This is what a jury does when it assembles evidence to reach a verdict (an agreed generalization) of guilty or not guilty.

The philosophical dialectic is a variant expression of the primary algorithm which has come down to us as something of an evolutionary product in philosophical thought. The pattern is expressed differently in Western and Eastern systems of thought, but they both represent attempts to articulate the intuitively perceived biological pattern. In fact, the history of the variants in the development of the philosophical dialectic can be seen as a sort of intellectual groping toward the most suitable expression of the primary algorithm.

This section will treat the West. The next will examine the East.

THE WEST

The philosophical dialectic in the West eventually came to be expressed in the familiar and now classical triadic pattern of:

THESIS
SYNTHESIS→ / ***→SYNTHESIS***
ANTITHESIS

It took a long time for this pattern to emerge clearly in philosophy. In fact, it was not until the close of the 18th Century that it was given the explicit triadic form by the German philosopher, Johann Gottlieb Fichte.[49] When we compare the triadic philosophical pattern to the pattern of physics and biology

ORDER → CHAOS → ORDER
INCORPORATION → EXTENSION → INCORPORATION

we quickly see that synthesis--the making of a unit, the creation of a whole out of parts--is a synonym for incorporation and order. In keeping with the matrix-product-matrix synthesizing loop, when we put the philosophical dialectic to work, we know that the process begins explicitly or implicitly with a matrix or synthesis. There must be an anchor, ultimately a somatic anchor, to any thought process. Even in thought, something is not created

[49]Fichte' s *Grundlage der gesamten Wissenschaftslehre* was published in 1794.

out of nothing. Therefore, at the beginning point, we are in synthesis...in the vernacular, at this point, "we've got it all together."

Then, something happens. Gratuitously or otherwise, a contradiction, something not included in the first synthesis, appears. The beginning synthesis is broken and we enter the extension or chaotic phase. The old synthesis, still anchored to its somatic matrix, now becomes the thesis, the position to be defended. The contradictory item, which has broken the previous order, becomes the antithesis. The challenge is to achieve a new synthesis (incorporation or ordering) that contains and supersedes both the thesis and the antithesis. When this is done, we again have our act together...everything fits. The process continues on and on, on and on.

Since the underlying pattern is an energized or motorized bioneurological one, we don't need to wait for contradictions to appear in the external environment The motorized algorithm assures that as long as we are alive and functioning we will be making discriminations, seeking contradictions or environmental items not included in previous orderings or incorporations by extending ourselves into the environment. The nature of our neural network based cognitive functioning is to generate its own contradictions, make its own discriminations, as a matter of function.

Again, we are irresistibly seekers, explorers of our environment. This clearly survival mechanism is greatly amplified in the massive neural network of the brain. We seek to reach into, to discriminate, then to order, incorporate, synthesize our environment as a way to control it...to assure the maintenance of our vital processes...our very survival. The process is anchored through the three dimension protein somatic structure ultimately to the genetic informational matrix.

The philosophical dialectic sounds forbidding to many nonphilosophers. But it does not have to be applied only to lofty and abstruse philosophical questions. A simple example can make the workings very clear. Suppose we begin, again, arbitrarily in synthesis. We have identified an object in the environment that has certain characteristics and have given it the name ***woman.*** Right now that's all we see in the environment and we understand it. We are in synthesis.

But wait! Suddenly we see (discriminate) something else out there. And it's different. We don't know what it is, but it's ***not woman***. Our synthesis is broken. We no longer fully understand our environment. It's chaotic or disordered. We're impelled to do something about it. It could threaten our survival.

Woman, now becomes the ***thesis***. The ***not woman***, the result of our perception, discrimination or reaching out, becomes the ***antithesis***, the contradiction. We study the ***not woman*** and because of its different characteristics decide to call it ***man***. Since ***woman*** and ***man*** are ***not*** the same thing, we must create a new synthesis which includes them both. We do this by moving to a higher level of generalization (a synonym for synthesis). We could call this new synthesis, which relates the two items, ***human being***. Graphically, the process can be represented as follows:

1) ***Woman***

2) ***Woman***
Not woman

Woman
3) ***Woman →*** → ?
Not woman

Woman
4) ***Woman →*** → ?
Man

Woman
5) ***Woman →*** ***→ Human being***
Man

The process is, of course, reversible, as Piaget said in his description of intelligence. We can retrace the pathways of the neural network. That is, we can analyze the synthesis of human being into its component parts or lower level syntheses of woman and man. Since this rhythmic algorithm goes on as long as we live, we can't achieve ultimate synthesis, or get it all together, or arrive once and for all--except when the process itself stops...in death.

For this reason the philosophical quest for ultimate truth, so characteristic of Western thought, may be understandable, but unattainable. Small truths or facts may be ascertained, but as long as the organic machinery runs, humankind may well be haunted in its search for final truth by the rhythmic, relentless process of discriminating, self-contradiction, questioning, or doubt...and uncertainty.

The nature of the cognitive algorithm is that man will, in the act of extension, always pose the opposite possibility to any temporarily assumed ultimate truth. The ambiguity of the quest can be captured in the same algorithmic pattern:

ULTIMATE TRUTH
ULTIMATE TRUTH → / ***→ ????***
NO ULTIMATE TRUTH

or the reverse,

NO ULTIMATE TRUTH
NO ULTIMATE TRUTH → / ***→ ????***
ULTIMATE TRUTH

This brings us to another important point that we need to consider before getting into the history of the philosophical dialectic--the distinction between process and content.

PROCESS AND CONTENT

The key point to hang on to in the pages that follow is that the focus is on process, not content. Algorithms concern process. The distinction is both fundamental and important. Without grasping the difference, it becomes difficult to clearly see the movement of philosophical thought. The primary cognitive bioneurological dialectic is the algorithmic process of life and thought. It is like a plowshare cutting through the contingencies of the environment. The product that is turned up is content. Content is the result or product of past process. This does not suggest that content is not important. It is. But it is not the dynamic part of life; it is the static part.

If we keep in mind that the primary algorithm, whether clearly distinguishable or not, is the dynamic process of all thought, we can see how each thinker used it, misused it, interpreted it or misinterpreted it. We can also see that content is the product of a thinker's cognitive algorithmic process and has no inherent sanctity. We may accept or reject it at will, or based on new evidence, and go on to create our own content. We can reject content, but we cannot reject the algorithmic process.

Content can be infinitely varied, the algorithmic process is essentially the same. For this reason focusing on content can result in endless analysis, reinterpretation, redefinition, and confusion. And confusion is compounded if process and content are not distinguished. Witness the enormous scholarly literature dedicated to interpreting and reinterpreting any major thinker in history.

In summary, then, my position is that the primary algorithmic process provides the fundamental key to an understanding of the development of philosophical thought. In the overall view, process is primary, content is secondary. Our cognitive algorithm rooted in physics and neuroscience moves relentlessly on, reorganizing content over and over and over again. It deserves our primary attention. In fact, process and content also express the primary algorithm in the form:

Content →Process →Content.

From the global viewpoint, it is a flowing or process algorithm.

DIALECTICS IN THE WEST

This brief survey of Western philosophy will focus only on the definition and uses of the dialectic. And I'll deal with only a few towering figures who not only self-consciously thought dialectically, but contributed to the definition and application of that process in the literature of thought. All thinkers, however, think dialectically. Some don't know it. Some deny it.

Some do so consciously; others unconsciously. There is no other way to think.

Dialectical thinking has not always been called dialectics in Western thought. And when the term dialectics is used, sometimes it is employed narrowly, sometimes broadly.[50] As we look at some different "types" of reasoning, whether called dialectic or not, I will point out their dialectical nature.

ZENO AND THE ELEATICS

The Eleatic School was founded in the 5th Century BC by the Greek philosopher, Parmenides, at Elea, a Greek colony on the Italian peninsula. Zeno, a follower of Parmenides, may be thought of as the first self-conscious Western dialectician, at least Aristotle thought him so. Although his writings come down to us only through the writings of later philosophers, he is given credit for arguing for the unity of being by holding that the belief in the reality of change leads to logical paradoxes or contradictions. That scheme of things can readily be put into the later evolved triadic pattern of the philosophical version of the dialectical algorithm.

	UNITY OF BEING ***(Thesis)***	
UNITY OF BEING → ***(Synthesis)***	/	→ ***????***
	CHANGE ***(Antithesis)***	

Zeno's paradoxes became famous intellectual puzzles for future philosophers (Ryle 1954, Grunbaum 1967).

HERACLITUS OF EPHESUS

Also, among the early Greek philosophers, who had notions of dialectic was Heraclitus. He was approximately contemporary, if not slightly predating, Zeno. Heraclitus apparently resided, not in Italy or Greece proper, but in the Ionian city of Ephesus in Asia Minor.

Heraclitus saw change as taking place between opposites which represented a unity. In this sense he differed from Zeno who saw change as illusory or false. Heraclitus also introduced the concept of flux which he illustrated by metaphor of a river, which constantly flows and constantly changes. This concept of flux was interpreted by some as suggesting a

[50]For some of the ways the word *dialectic* is used in philosophical writing, see Roland Hall (1967: 385).

process of becoming, which was later to figure in the dialectic of Plato, and the modern dialectical philosophers (Kirk 1954, Wheelwright 1959).

The pre-Socratic Sophists also used what was called a process of dialectics in the sense of a dialogue, argument, or discussion between protagonists. However, they fell somewhat into disrepute because the largely rhetorical method became corrupted by some into a device for winning arguments rather than a true search for knowledge or truth. This is a criticism that is leveled at our dialogical dialectical, adversarial legal system in the United States today, when the objective for each attorney becomes to "win" a case rather than find out the truth.

SOCRATES AND PLATO

Socrates, known to us mainly through the writings of Plato, skillfully used the question and answer dialogue or dialectical methodology to expose faulty thinking and to advance knowledge. He criticized the Sophists whom he felt used it deceptively for the sake of winning arguments, thereby corrupting or trivializing the search for truth.

But it was Plato, who developed the dialectic as a method of inquiry. Plato (427-347 BC) was one of the intellectual giants of ancient Greece. In his early dialogues Plato followed the question and answer method believed to be typical of the historical Socrates. In later dialogues, however, he changes his conception to an emphasis on division as a method. Division involves a repeated analysis of more general concepts or schemes into less general ones. Or said another way, it is a process of breaking wholes down into their parts. It can also proceed in the opposite direction of gathering parts into wholes, or moving from the less general to the more general. This method of division is the beginning of our modern systems of taxonomy or classification.[51]

It is easy to see components of the primary algorithm in Plato's method of division. Wholes or generalizations are synonymous with syntheses or incorporations or order. Wholes become parts of greater wholes which, in turn, can be divided, or discriminated, again into lesser wholes. The fact that the process goes in both directions is not a problem. Reversibility, as we have seen, is a feature of the cognitive algorithm. Plato's dialectic is said to have an ascending and descending path. This is simply a matter of semantics. The pattern can be laid out in any direction, but it still remains the same. Although it can run in reverse (discriminate or take apart rather than synthesize), Plato's dialectic has an overall forward direction, a

[51]For an analysis of Plato's method, consult Sayre (1969). Plato's method of division can be seen in the dialogues Phaedrus and Sophist (Jowett 1937).

progressive movement through the realm of ideas toward the first principle of the whole, or ultimate truth and unity.

Plato's dialectic functioned in the realm of ideas. The distinction between ideas and physical reality is blurred in Plato's thought, but if anything, reality was viewed as the reflection of ideas rather than the source of them. The dialectic path ascended from the lower level of reality to the higher and predominant level of ideas.[52]

The tendency toward a forward, progressive movement in Plato's dialectic leads us to describe it as a dialectic of realization, of becoming. Plato's dialectic moves in a sort of "jaws" effect to incorporate more and more concepts and ideas until it reaches toward an ordering or unity of all things. This jaws effect is significant because it sets the basis for what seems to be a fundamental distinguishing characteristic between Western and Eastern philosophy and mysticism which I will return to later.

ARISTOTLE AND THE SYLLOGISM

Aristotle (384-322 BC), the most famous student of Plato, moved Plato's question and answer method of dialectic from the peak of science and philosophy (where Plato had positioned it) to its base. In its place he put his formal system of logic based upon the syllogism (Adler and Gorman 1952: 348). However, the syllogism itself was apparently based on Plato's dialectical method of division (Roland Hall 1967: 378).

The syllogism is a deductive method of argument, which means that it proceeds from the general to the specific, from a higher level of generalization to a lower level. Or to put it another way, it proceeds from a whole to its parts. Aristotle also recognized induction, the complement, or reverse of the syllogism.[53] The difference between Aristotle's logic and Plato's dialectic is as follows:

In the method of logic, inquiry is carried on without dialogue. It proceeds from "true and self-evident" premises according to formal rules. Plato's dialectic, however, proceeded by way of dialogue from premises based upon generally accepted opinion. A classic example of the logical syllogism is:

> All men are mortal.
> Socrates is a man.
> Socrates is mortal.

[52]Gorman and Adler (1952: see esp. p 347). Plato's *Republic* Book IV and his dialogues *Sophist* and *Statesman* give examples of the ascent from appearance to reality and the ascending and descending paths of Plato's dialectic.

[53]The earliest system of formal logic is Aristotle's syllogistic method, developed in his *Analytics*.

As can be seen, progression is from the general to the specific. The deductive process can be graphically represented as follows:

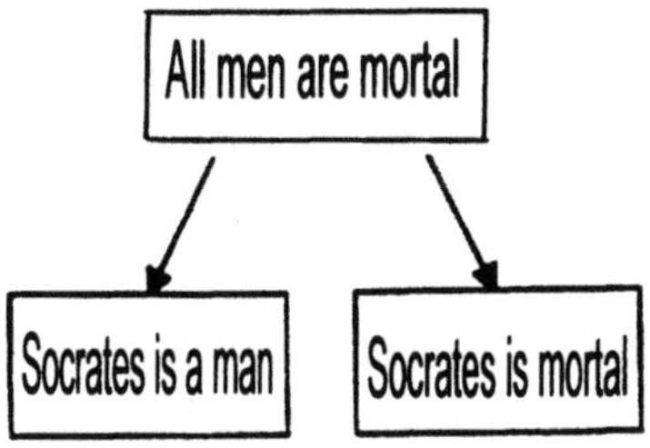

Figure. 8-1. The direction of deduction.

The top box represents the highest level of generalization or synthesis; e.g., *all men are mortal.* The two deductions, or parts that are made, are *Socrates is a man* and, therefore, *Socrates is mortal.* This was proved, of course beyond a shadow of a doubt when Socrates drank the hemlock and died.

Induction works in exactly the opposite direction. However, in most syllogisms a simple doing it backwards won't work. That is, we can't put the components *Socrates is a man* and *Socrates is mortal* together and come out with the synthesis that All men are mortal. But the major premise, *All men are mortal*, is a "self-evident truth" or generalization arrived at originally by inductive process based upon the observation of many men. And then, it was used deductively to establish that one object, Socrates, being identified as a man, could logically be considered to have a characteristic of all other men...mortality.

The inductive process, then, can be graphically represented with three blocks and arrows flowing in the opposite direction...from components to generalization.

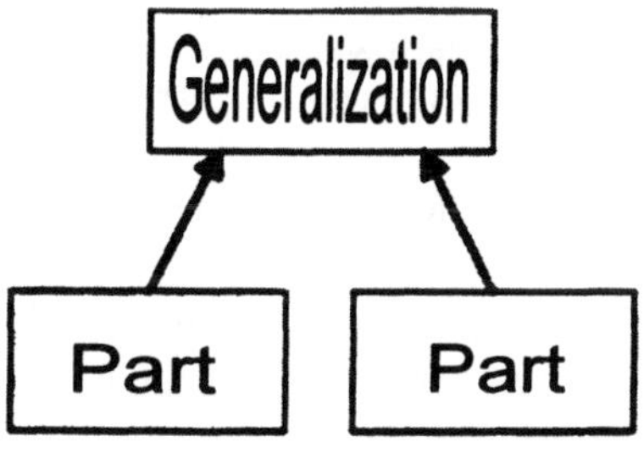

Figure 8-2. The direction of induction.

If we put the two processes together, they are shown to be mirror images of the same pattern. Assume there are only two men in the world and follow the diagram below.

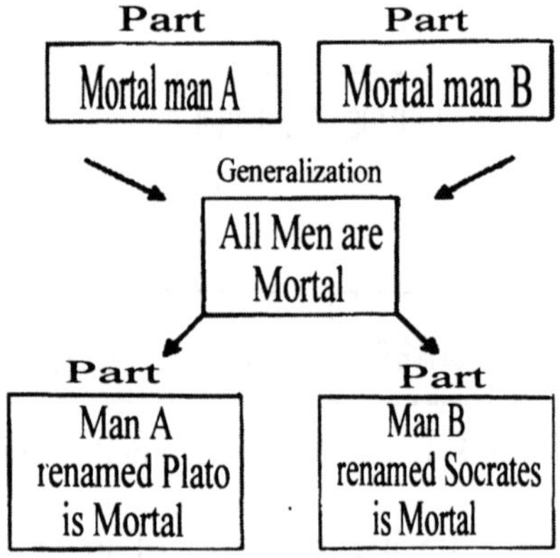

Figure 8-3. Deduction and induction as mirror images showing their inherent identity.

Assuming there are only two men in the world may make the illustration seem a little trite, but it serves to demonstrate the basic identity of the patterns. If we now turn the pattern on its side and give movement to it, we can see that it matches the pattern of the underlying primary algorithmic dialectic.

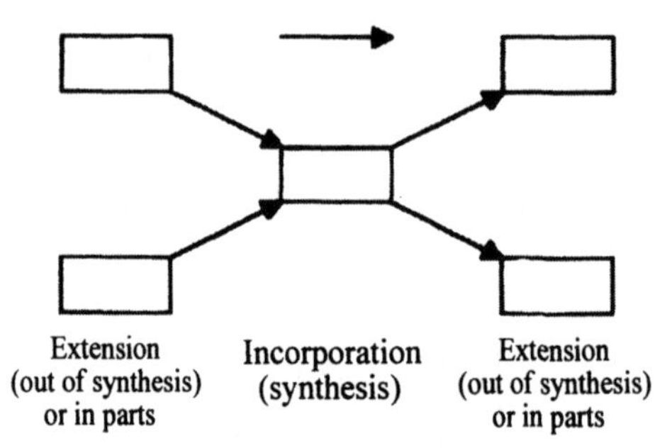

Figure 8-4. The dialectical flow of the primary algorithm.

Although Aristotle didn't call his formal logic *dialectic* because the term meant other things to him, we can see the primary algorithmic features of deductive and inductive reasoning.

Later, the discipline of logic was divided into deductive and inductive branches. All forms of argument and analysis can ultimately be included under the umbrella of either deduction or induction...the assembling of parts into wholes or disassembling of wholes into parts, putting things together or taking them apart.[54] Since deduction and induction are patterns of the primary algorithm and all forms of argument and analysis are either deductive or inductive, it follows that all forms of argument and analysis are patterns of the organic dialectic.

The discipline of logic therefore has its basis in biological functioning and an understanding of this biological functioning has allowed us to grasp the basic organizing patterns of logic. Deduction and induction, as formal methods can perhaps be thought as freeze-frames of the moving pattern of the dialectic held in place for purposes of analysis.

DIALECTIC IN THE MIDDLE AGES, OR HOW MANY ANGELS CAN DANCE ON THE HEAD OF A PIN?

The dialectic as a method of inquiry fell into disrepute in the Middle Ages because of its use by philosophers and theologians to reason endlessly without reference to fact or experience. The dialectician was depicted as one who argues rather than observes and who by reasoning without reference to reality often pushes premises to the point of absurdity.

English philosopher Francis Bacon (1561-1626) branded scholastic philosophy, because of its empty dialectics, as containing little substance and "infinite agitation of wit"(see Adler and Gorman 1952: 345). Medieval scholars simply turned on their cognition with no reference to fact and went endlessly off into abstractions.

This points up an important feature of the cognitive algorithmic functioning. It operates on what it is fed. This means that if you put in quality, you tend to produce quality. If you put in garbage, you tend to produce garbage. This, of course, is much like our modern computers. And, if you do not tie the cognitive function to fact and experience or empirical reality, it will run off into abstraction, fantasy, and absurdity.

DESCARTES AND THE RENAISSANCE

Rene Descartes (1596-1650), the greatest of French Renaissance philosophers, criticized the dialectical method to include the syllogism as incapable of generating premises or contributing to the discovery of truth. This is because the formal method as such could proceed only after being given an initial premise to work with. This is similar to the way a computer would work. It must be fed basic premises or instructions to produce anything of value. If not fed it doesn't produce anything. The difference, however, is that the cognitive process is always running and, because of the

[54]Beardsley (1975) is a well-written book on logical thinking. See the section on induction and deduction. Also see Dewey (1933). Palmquist (1995) takes an interesting approach in his introductory text.

nature of cognitive structure, there will always be something there to work on.

The formalized dialectical method that Descartes was criticizing was a limited, externalized and somewhat distorted reflection of the biological dialectic process. Descartes believed that initial premises could not come from the dialectical method then being practiced, but rather had to come from intuition (or subjectively) and then be treated deductively and checked by analysis of experience and investigation.[55]

In his method, Descartes was actually closer to the basic algorithmic process. He realized it could not operate productively in a vacuum and so he brought it back into the head and hooked it up to subjective experience (i.e., the somatic matrix) from which it must then draw its initial premises. He added, of course, the additional important requirement of checking the process with empirical reality.

KANT AND THE TRANSCENDENTAL DIALECTIC

Plato saw the basic unity of the world of ideas and the world of reflected reality, with the dialectic ascending from the latter to the former. Immanuel Kant (1724-1804), the great 18th Century German philosopher, shattered the integrity of that unity. Kant claimed that delusion and error result when pure reason tries to apply the ideas dialectically to objects of the tangible, phenomenal world, and then falsely believed the objects to correspond to the ideas. Kant asserted that pure reason, detached from the experiential world, functioned dialectically in such a manner that any conclusion was, as a matter of function, automatically opposed by an opposite conclusion equally acceptable to reason.

Because of this inevitable balance of reason against itself, Kant held that the dialectic of pure reason was illusory. Kant considered the dialectical illusion a natural and therefore unavoidable characteristic of human reason based upon subjective principles which were imposed upon us as objective by the mind. The illusion was particularly obvious when the mind tried to transcend experience and answer impossible questions.[56] Kant's position reminds us that we can't leave the cognitive machine running unharnessed to experience and expect anything useful to come of it. Whereas Plato always saw the dialectic leading toward unity or synthesis, for Kant it always ended in contradiction, antithesis, extension, or chaos.

[55]See the section on Descartes in Encylopedia Britannica (1970, v.7: 282. Also consult Descartes *Rules (III and X)* in Hutchins and Adler, V. 31: 4, 16-17.

[56]See the discussion of Kant's treatment of the dialectic in Adler and Gorman (1952:. 348-349). Consult Kant's *Critique of Pure Reason* (1969), trans. by K. Smith.

HEGEL AND THE MYSTIFICATION OF THE DIALECTIC

Georg W. F. Hegel (1770-1831), another eminent German philosopher, in reaction to Kant, restored the unity of the world of ideas and the world of reality. Hegel saw the dialectic as the process of mind, or the Idea, in ceaseless motion toward absolute truth. But it was more than that. It was the working of Absolute Mind, or the Idea, or spirit, toward progressive realization or objectification in history. Hegel, thus, moved the dialectic out of the realm of human cognitive activity and assigned it a cosmic source and a historical mission. Hegel saw the dialectic as moving progressively by the resolution of contradictions to higher and higher levels of synthesis.[57]

MARX AND THE EXTERNALIZATION OF THE DIALECTIC

Karl Marx (1818-1883), the father of modern communism, studied Hegel closely and felt that he was wrong. Marx maintained the unity of the world of ideas and world of reality, but reversed their standing. According to Marx, Hegel viewed the real world as only the external, phenomenal form of the Idea, or Absolute Mind. Marx, on the other hand, posed the contrary view that the ideal was no more "than the material world reflected by the human mind and translated into forms of thought."(see *Capital*, preface to the second edition 1906). Marx, thus, reconstructed Hegel's dialectical pattern of history into the conflict of material forces, rather than the progressive objectification of spirit. In Marx's opinion, he demystified Hegel. What he actually did was move the mystification out into the material world...a dubious improvement.

Marx saw his conflict of material forces exemplified in the class struggle. By his view of the dialectical movement of history toward the ideal social synthesis of communism, Marx's dialectic, like that of Plato and Hegel, was a dialectic of realization or becoming.

AND SO

Dialecticians abound in Western thought and are more plentiful now than ever, and so we can't review them all. I think, however, we have looked at some of the most prominent and influential, and that is sufficient to establish the main character of the Western dialectic. In summary here is what these thinkers did to the biological dialectic over the course of history:

Plato and Aristotle externalized it as a method. Plato did so in his dialectical question and answer method and in his method of division. Aristotle did so in his formal logic. Plato established the dialectic as a

[57]Loewenberg (1957) *Hegel Selections*. Discussions of Hegel's dialectic are in Roland Hall (1967) and in Adler and Gorman (1952:.349-350).

process of realization, becoming...a tending to an incorporation of everything.

Bacon, Descartes, and Kant criticized it as a method. Kant realized that it was the way reason functioned naturally.

Hegel and Marx both continued the Platonic tradition of forward movement and realization. Hegel assigned the dialectic a cosmic source in the form of Absolute Mind and also moved it into history as a shaping force. Marx assigned the source of the dialectic to the material world and saw it expressed in class struggle.

Marx and Hegel wreaked the most havoc on the dialectic because they took it out of the mind and body where it belongs and made it a spiritual, material, and historical force. They reified it. And in the process they mystified in a way that none of their predecessors had done.

To their credit Hegel and Marx did see that the dialectic has a shaping effect on society and history. They failed completely, however, to distinguish the first and second dialectical algorithms and to properly ground them. As a history-making or shaping phenomenon, it is the second, reciprocal algorithm or dialectic, the tug and pull of ego and empathy rooted in our triune modular brain structure, that we must look to.

By the way of protein-nucleic acid synthesis, the human organism is by its very nature in a fundamentally dialectical relationship with its environment. It is the reciprocal, second dialectic of our triune modular brain structure, however, which is the fundamental socially, and accordingly historically, interactive aspect of our human nature.

The behavioral tension that drives the reciprocal dialectic originates within each of us and becomes externalized between ourselves and others. It thereby creates the tension of inequities and the issues of freedom between and among us which further become formalized in our social and political structures. The tug and pull between ego and empathy within us, and as externalized, is a dynamic, shaping force in our social history. It is, however, not a simplistically, directional process. It is influenced at any moment in time by other variables which come into play when we shift our perspective to the disciplines of anthropology, sociology, economics, and political science. These are such variables as historical tradition, culture, population, resources, technology acted upon by political and economic structure and decision-making processes.

A full consideration and analysis of these could not be undertaken here but would be the subject of another book (e.g., see Cory 1999, for a partial effort). But the essential point to be made is that the algorithms act as continuing bridging variables among these disciplines which can't be properly understood without appreciation of the algorithmic dialectical processes, particularly that of the second, reciprocal algorithm, which actively underpins them.

The thinkers I have considered, however, had no adequate foundation for the dialectical process in physics, biology, and human behavior. They were, in effect, operating in the absence of an effective definition of human nature. One of our finest recent existentialist, dialectical thinkers, the French philosopher Jean Paul Sartre recognized and lamented this lack of definition and went on to do the best he could without it.[58]

On the other hand orthodox Marxists were usually not overly concerned with the absence of a definition of human nature, since they saw human nature and thought patterns as reflecting what they believed to be the laws of dialectic in nature. Marx's concept of the dialectical laws as residing in nature was expounded and developed further by Engels (1934) and followed by Lenin. This set the orthodoxy for Marxist-Leninist thinkers of the former Soviet Union and its satellites.

Nevertheless, some Soviet thinkers, especially those versed in physics, were troubled by the dogma of laws of dialectic in nature as spelled out by Engels and Lenin and questioned the universality of the official concept. For example, M.N. Rutkevich writing in Moscow in 1958[59] stated the dialectical formulation did not hold in microphysics. The challenge was supported by a Rumanian scholar, Calina Mare, in 1965 (Mare 1965, in Dumitriu 1977).

Such challenges to the universality or generality of the dialectical laws as existing in nature tended to undermine the theoretical basis of orthodox Marxist materialist dialectics, and the position of Marx himself. After all, in "standing Hegel on his head," Marx wrote that Hegel's ideal was "no more

[58]At several points in his writings, Sartre makes the point that there is no common human nature. At the close of *The Problem of Method*, in discussing the application of dialectic to anthropology, Sartre states that the task would be easy if some kind of human essence could be brought to light (see 1963: 169-70).

In *Critique of Dialectical Reason*, Sartre argues against Engel's dialectics of nature, which sees dialectics as laws of nature external to humans but acting on them historically. According to Sartre, this makes the dialectic dogmatic and leaves no room for the freedom and creativity of humans. He does not, however, deny the possibility of dialectical connections in nature, but does not believe at this point of our knowledge that we can either confirm or deny such a dialectic. He states that each of us is free to believe or not to believe that physico-chemical laws express themselves dialectically.

Although he is unable to source or ground the dialectic adequately, Sartre claims that it reveals itself in and through human activity or praxis and that therefore we should seek it where it presents itself in our social life, our social universe (see 1976: 29-36).

[59]Rutkevich (1958). referenced in Dumitriu (1977). See also the discussion of the status of dialectical philosophy as it developed more meaningful controversy in the last decades of the former Soviet Union (Scanlan, 1985:especially, 133-142).

than the material world reflected by the human mind and translated into forms of thought." If there are no dialectical laws in the material, nonorganic world, Marx was simply wrong.

My position is, of course, that there is ***no*** dialectic as we know and experience it in inanimate matter, but the dialectical algorithm first appears with the organic synthesizing loop and related metabolism of living organisms.

And, of course, none of the great thinkers of the West distinguished the fundamental subject/object relationship of the primary algorithmic dialectic from the society-building ego-empathy dynamic of the second, reciprocal dialectic.

Chapter 9

HOW WE PHILOSOPHIZE: DIALECTIC, BEING, AND HARMONY IN THE EAST

Do not pursue Being, nor get attached to Not-being.
Be ever in Oneness and the two will vanish by themselves.
(Sosan, Zen patriarch, 7th century) (Suzuki 1972: 65)

The Chinese apparently developed a dialectical concept during approximately the same historical period as the West...around 500 BC. This is noted by the German sinologist Hellmut Wilhelm as a notable instance of parallelism.[60] In our comparison with the West we'll look mainly at the philosophy of China indicating briefly its influence from ancient beginnings to the present day.

This early dialectical concept is the organizing theme of the *I Ching* or *The Book of Changes*, the fundamental book of Chinese philosophy, as it began to take shape. The origins of the *I Ching* are lost in antiquity, but elements of the book probably date back in its earlier forms to 2,000 years BC or earlier. The *I Ching* was primarily used as a book of divination and even today it is used that way.[61] Aside from that, however, and more importantly, it is a book of wisdom. And it is from this perspective that we are interested in it here. As a book of wisdom, it became the basis of the major streams of Chinese philosophy, Taoism and Confucianism.[62]

[60]Wilhelm (1960: 13), also Wheelwright (1959: 3). This time frame is not without dispute, however, see Chan (1963: 3-4).

[61]For some varied uses of the I Ching in the period 960-1279 AD see Smith, et al.(1990).

[62]For translations of the I Ching see Blofeld (1968) and Wilhelm (1967).

Underlying the I Ching is a dialectical concept of a primal unity of the cosmos within which context change or movement is perpetual. Confucius described it as like the ceaseless flow of a river. Lao-Tsu called it the tao, the immutable, eternal law at work in all change, the principle of the one in the many (Wilhelm 1967: liv-vi, 54-56).

Wilhelm points out that nonchange is the background for all changes. The world is conceived as a cosmos...a unity...rather than a chaos. "This belief," he writes, "is the foundation of Chinese philosophy...the ultimate frame of reference for all that changes is the nonchanging."(Wilhelm 1967: 280-281).

The dialectic is expressed in the *I Ching* by the symbol

which represents the t'ai chi t'u, or primal beginning. The light and dark sides of the circle, called the yang and the yin respectively, gradually came to represent the two polar forces of the cosmos, the positive and the negative. The line at which they meet, originally called the ridgepole, represents oneness--the union of opposites. The dynamics of the dialectic--the constant movement or change--derives from the tension between the two primal forces, which arises again and again causing them to unite and be regenerated.

R. G. H. Sui sees it as a pattern of order turning to chaos to order again... to be repeated forever and ever (1968: [406] 64). This is, of course, the first dialectical pattern of the primary algorithm projected upon the cosmos.

Since the dialectic of the *I Ching* starts with cosmic unity and undergoes chaos within the context of that unity, we may express its characteristic dialectical pattern as a variant of the primary algorithm as follows:

CHANGE
COSMIC UNITY → (increase/decrease of multiplicity caused by tension of opposites) →COSMIC UNITY

The dialectic of the yin-yang thus involves ceaseless change within the overall context of unity. It has a feature of returning or reversal about it. This feature seems to be expressed more clearly in the *Tao Te Ching*, or *The Canon of the Way and Its Attainment,* attributed to the Taoist philosopher, Lao-Tsu (Sixth Century BC) (Chang 1975: x-xi, 48).

The dialectic of Lao-Tzu, in contrast to that of the West (especially that of Hegel), has no movement toward an all-encompassing unity of all. It is not a progression from a synthesis of lower level and lesser content to a synthesis of higher level and more content. In short, it is not a dialectic of realization or becoming. Rather, it is a dialectic of the self-identity of contradictions. Such opposites as being and non-being, goodness and evil, etc., are found within themselves rather than resolved in a higher synthesis (see the discussion in Chang 1975: 9-10). The way is also called the way of harmony, in which contradictions are resolved, heaven and earth made one (e.g., see the translation in Cleary 1991: 10).

Lao-Tzu states that the man of Tao is characterized by the attainment of quiescence. Quiescence is gained by reverse movement or a returning, an unlearning. At one point Lao Tzu writes:

> To learn, one accumulates day by day
> To study Tao, one reduces day by day
> (Chang 1975: xi, 131-132).

The dialectic of the Tao, thus seems to be characterized by learning and unlearning, increase of multiplicity and decrease of multiplicity, becoming and not becoming. But to achieve the underlying unity, one must return or move in reverse. Tao must be achieved by non-conceptualization (Chang 1975: xvii).

This enlightenment by non-conceptualization is another contrasting feature with the Western dialectic.

In the West, Plato conceived the realm of ideas to be the only reality, and this reality, consisting of a logical system of ideas, could be grasped only by the highest faculty of reason. That is, it required the highest level of conceptualization.

On the other hand, the reality of the Tao, is formless and can only be grasped directly and spontaneously through intuition. Therefore, conceptualization, or discursive reasoning, is more than useless in achieving the Tao, it gets in the way (Chang 1975: xii). The reverse motion, unlearning, or achievement of enlightenment by non-conceptualization, is a significant taoistic feature of Chan or Zen Buddhism (Wilhelm 1960: 33-34; Linssen 1972: 59-64; Van Over 1973: vii-xxiv).

According to D. T. Suzuki, a prominent Zen scholar, the main philosophical point of Zen study is reflected in the words of the Zen patriarch, Sosan(7th century):

> Do not pursue Being, nor get attached to Not-being.
> Be ever in Oneness and the two will vanish by themselves
> (Suzuki 1972: 65).

BETWEEN WEST AND EAST

In summarizing the contrast between the dialectic of the West and that of the East (as represented by the Chinese), we can with some accuracy refer to the former as a dialectic of realization or becoming and the latter as a dialectic of recognition. In the West, the dialectic begins at the point of smallest content and proceeds to a synthesis that incorporates all of reality. In the East, the dialectic involves a recognition or perception of the underlying unity of all things. What is central and germane is the demonstration that both patterns can be identified with the basic underlying biological primary algorithmic pattern. They are:

	THESIS	
SYNTHESIS→	***/***	***→SYNTHESIS***
	ANTITHESIS	

	CHANGE	
COSMIC UNITY→	***(increase/decrease of multiplicity caused by tension of opposites)***	***→COSMIC UNITY***

The philosophical dialectic offers a good point of entry into the subject of mysticism, which I will consider in the next chapter.[63]

[63]For an attempt to reconcile the I Ching and the teachings of Jesus, which is some sense falls between philosophy and mysticism, see Lee (1976).

Chapter 10

BRAIN ALGORITHMS IN MYSTICISM, CANNIBALISM

Thy voice from on high crying unto me, "I am the Food of the fullgrown; grow, and then thou shalt feed on Me. Nor shalt thou change Me into thy substance as thou changest the food of thy flesh, but thou shalt be changed into Mine."

(St. Augustine, Confessions, Bk. vii)

Talking about the dialectic in philosophy leads easily off into mysticism. In fact, it's sometimes hard to separate the two since they tend to flow into each other where cognition and subjective experience merge.

Pure logic and its mathematical and symbolic offshoots employ chiefly the primary algorithm. Cognition is running, seemingly remotely connected to its somatic matrix and detached from the second, reciprocal algorithm of the triune modular brain structure--the tug and pull between ego and empathy driven by behavioral tension. In the more subjective areas of philosophy, ethics, and perhaps esthetics, the first and second algorithms, however, tend to merge and become confounded. In mysticism, and especially in Western mysticism, the two algorithms become so merged and confounded that the pure logician throws up his hands in despair at the contamination by subjectivity.

Said differently, philosophy and mysticism tend to flow into each other where the first and reciprocal algorithms are merged. This distinction has never before been effectively made in either philosophy or mysticism. And that has lead to a great deal of muddled thinking. In going through the discussion of mysticism, every effort will be made to achieve this clarification. The first step is a definition.

WHAT IS MYSTICISM?

Mystical experience is the subjective perception of the functioning of the algorithmic processes given a transcendental or supernatural interpretation. The particular quality of the mystic experience depends on three factors. They are:

(1) the cognitive content given it,

(2) the incorporative mode of the reciprocal algorithm (determined by the tug and pull of ego and empathy),

(3) whether the experience is one of amplification or diminution.

Since the terms *amplification* and *diminution* have not been used before, they need some additional definition.

Amplification and diminution go back to brain structure and the variable of activity level which was described as working together with ego and empathy in the reciprocal algorithm. The amplifying qualities of the human brain have often been noted by scholars of neuroscience and are inherent features of protein function (e.g., see Girault and Greengard 1999: 37; Bray 1995) even before the development of specialized neurons and neural networks. Amplifying can be turned up or down. When we turn it up, we have amplification. When we turn it down, we have diminution. Amplification and diminution can be thought of as the subjective equivalents of high activity level and low activity level.

In the next few pages, I will develop further this definition of mystical experience and support it with materials from scholarly and mystical literature.

HOW IS MYSTICISM ALGORITHMIC?

The first job, then, is to say why mystical experience is expressed, and apparently experienced as well, in terms of our brain algorithms. There are two reasons for saying so.

First, the primary algorithmic features of extension and incorporation are basic defining characteristics of mysticism all over the world. Everywhere the mystic becomes extended and unified (incorporated) into a greater whole, usually identified as a deity, cosmic spirit, or cosmic mind.

Secondly, the mystical literature indicates clearly the connection between eating or physical incorporation and the achievement of mystical union (incorporation).

A review of selected materials will demonstrate both of these claims.

MYSTICISM AND EXTENSION & INCORPORATION

First, I have claimed that mystical experience involves the algorithmic features of extension and incorporation--the achievement of a greater unity. Evelyn Underhill, Oxford lecturer in the philosophy of religion, in her classic study, *Mysticism* (first published in 1911), defines mysticism in such terms. Underhill writes:

> mysticism, in its pure form, is the science of ultimates, the science of union with the Absolute, and nothing else....(Underhill 1961: 72).

She also writes of "the deep instinct of the human mind that there must be a unity, an orderly plan of the universe"(Underhill 1961: 29). And she sees that union, a becoming incorporated with the "Absolute Life" as "the true goal of the mystic quest"(Underhill 1961: 170).

Professor W.T. Stace of Princeton University, in a more recent work, *Mysticism and Philosophy* (1960), analyzes mysticism worldwide and divides it into two broad classes, extrovertive and introvertive.

Extrovertive mysticism emphasizes "the Unifying Vision--all things are one."

Introvertive mysticism emphasizes "the Unitary Consciousness: the One, the Void; pure consciousness."[64]

Both categories involve a total incorporation.

The connection between eating or physical incorporation and the achievement of mystical union is especially apparent in the Christian tradition. The central sacrament, the Eucharist, Mass, or Holy Communion is structured around this connection. The participants eat the bread and drink the wine which symbolize the body and blood of Christ. They do this in order that they may be

> ....filled with thy grace and heavenly benediction, and made one body with Him, that He may dwell in us, and we in Him.

Or similarly, that they may become

> ...members incorporate in the mystical body of Thy Son, which is the blessed company of all faithful people...[65]

In the *Gospel according to St. John*, Jesus is recorded to say:

> For My flesh is true food, and My blood is true drink.
> He who eats My flesh and drinks My blood abides in Me,
> and I in him (55-56).

This theme appears frequently in Christian mysticism sometimes not tied to reference to the Eucharist. For example, St. Augustine writes in his *Confessions*:

> Thy voice from on high crying unto me, "I am the Food of the fullgrown; grow, and then thou shalt feed on Me. Nor shalt thou change Me into thy substance as thou changest the food of thy flesh, but thou shalt be changed into Mine."[66]

[64]Stace (1960: 131-132). Stace analyzes Christian, Islamic, Jewish, Hindu, and Mahayana Buddhist texts. The last greatly influenced Chan Buddhism of China and Zen of Japan. Also see Van Over (1973) who draws upon Stace in a study of Chinese mystics, and Naranjo and Ornstein (1971: 35).

[65] *The Book of Common Prayer*, Protestant Episcopal Church (1945: 81-83).

[66] Augustine, *Confessions*, Book vii., cap. x (quoted in Underhill 1961: 419).

This statement by Augustine clearly shows a perception of the difference between physical metabolism and the cognitive algorithm. The eater does not change the substance into his own body, his body is changed into that of the Deity.

And Ruysbroeck, a great mystic of the Middle Ages exclaims:

To eat and be eaten! This is union.... (quoted in Underhill 1961: 425).

From this statement we can't be sure that Ruysbroeck gets the distinction, or whether he is just expressing himself poetically and joyfully.

Similar references can be found in other mystical traditions. The ancient *Upanishads*, chief theological documents of the ancient Hindu religion, identified Brahman, the ultimate ground of the universe, with the process of eating and being eaten (Zaehner 1957: 136).

More recently, the Swami Yogananda, in his *Autobiography of a Yogi*, describes the mystical experience of samadhi as incorporating to oneself all things. In contrast to St. Augustine, he tells us that he "swallowed" and "transmuted" every particle of universal dust, even good and bad, into the ocean of his own being (Yogananda 1994: 145-146). Such a total, devouring, incorporating to oneself, not only indicates a confusion of the physical and cognitive, but also parallels Nietzsche's concept of the will-to-power.

MYSTICISM AND CANNIBALISM

The clearest connection between eating and mysticism is to be found in the literature on sacrifice and cannibalism. There seems to be a direct link between mystical experience and cannibalism through the medium of religious ritual. Although the evidence may never be conclusive, it seems a possibility that primitive humanity practiced cannibalism--perhaps universally.[67]

Of special interest is the fact that at several archeological sites, skulls seem to get special attention, as if they were broken open and the brains removed for sacramental consumption. If this is true, early cannibalism at this point already had a cognitive element...the desire to incorporate attributes or powers, rather than merely an act of physical nutrition.

Of comparative interest, the history of cannibalism among primitive peoples shows such cognitive features. People eat others to incorporate admired attributes, to achieve control over the events of their lives, to gain power, commune with the dead, and to commune with the gods. Such factors predominate over the desire for food.

Eli Sagan, gives us a scholarly and comprehensive work of special interest to us on this subject in his book titled *Cannibalism* (1974). Prior to Sagan's

---

[67] See Arens (1979, 1998) for a counterargument that sees cannibalism as largely a matter of myth.

analysis the subject was somewhat taboo for book length treatment. Since then, however, there have been a number of serious book-length studies. From somewhat taboo status cannibalism has become a favorite theme among anthropologists, not only because it is important, but probably because it lends excitement to the discipline.

Marvin Harris's well known *Cannibals and Kings* (1977) puts forth the thesis that the Aztecs practiced cannibalism for the protein rather than for mystical or symbolic reasons. This interpretation has been amply criticized as narrowly materialistic by other prominent scholars. I. M. Lewis of the London School of Economics and Political Science has gone so far as to call it absurd (Lewis 1986: 77).

Marshall Sahlins, a well-known anthropologist, states : "Cannibalism is always 'symbolic' even when it is real"(Sahlins 1983: 88). Peggy Sanday's recent work, *Divine Hunger* (1986), also sees symbolism and religion as central to the practice. Key Ray Chong (1990) has reviewed the literature and reported on extensive cannibalism in the history of China.

Sagan, however, ventured to give us an especially relevant and interesting developmental sequence in the ritual of sacrifice, which we may interpret and connect with the evolution of mysticism. This sequence is summarized as follows:

1) *Cannibalism with no idea of sacrifice.*

This stage characteristically involved the incorporation of cognitive, or non-material elements; such as the attributes of the deceased, maintaining the presence of the deceased (for instance, a family member), increase of power, etc.

2) *Humans sacrificed and their flesh eaten.*

This stage involved the progression to an abstract concept of a deity, or force of nature. By having the gods participate, their good-will, blessing, neutrality, etc., were incorporated by their sharing in the cannibalistic meal. This was an obvious upward valuation of human life over the first stage.

3) *Humans sacrificed but not eaten.*

This represents a yet higher stage in which humans are sacrificed as food for gods, but not for human consumption. Again, the cognitive elements are much the same as the second stage.

4) *Animals sacrificed in place of humans.*

This stage exemplified in the Biblical literature by the story of Abraham and Isaac in which Abraham was willing to sacrifice his son, Isaac to God, but was at the last minute told by an angel that the a lamb should be substituted. This stage shows a further progression in the valuation of human life with approximately the same cognitive elements as stages two and three.

5) *The symbolic act of the crucifixion, and the eucharist.*

This stage represents a further elevated valuation of human life and an identification of it with the deity. It includes the cognitive incorporation of humans with the godhead.

Each of the five stages offered by Sagan represents an increasing level of cognitive discrimination. Sagan, arguing from a Freudian position, sees this sequence as a progressive sublimation of aggression (Sagan 1974: 59-63). From the standpoint of the bioneurological algorithms, I would say that the stages represent progressive discrimination of metabolic functioning at increasingly higher levels of cognition--that is, from the basic level of physical incorporation to higher levels of cognitive function.

I might even suggest a sixth stage to Sagan's list--that of a mystical incorporation with a deity, absolute, or cosmic source without the vestigial physical remains of the bread and wine. This would be the totally abstract, cognitive incorporation with the absolute characterized by some of the world's mystical religions and philosophical systems.

So we have our connection between eating and mysticism--the incorporation of physical materials, and the incorporation of the non-physical, such as attributes, concepts, deities.

ALGORITHMIC DISCRIMINATION AND THE DEVELOPMENT OF HIGHER CONSCIOUSNESS

Primitive man ate his fellow man, at least in part, because he could not distinguish the difference--that is, the impossibility of incorporating cognitive elements physically. He was functioning on an analogical basis which he confused with reality. It was, however, the metabolic character of his cognitive function that seemingly drove him to it.

Progressively, as he developed the capacity to distinguish his physical functioning from his cognitive algorithms, he moved from cannibalism to the higher levels of mysticism.

Contrary to what Freud thought, it was likely not a process of sublimation, but of increasing cognitive representation and discrimination.

The main point is that modern discriminating humans are not necessarily latent cannibals, waiting to, or in danger of reverting to savage practice, as the concept of sublimation might indicate. But, the cannibalistic urges which any child or adult may experience, may well be based on a confusion of, or failure to discriminate maintenance from cognitive levels of function. In fact, the development of higher consciousness among human societies can, from this perspective can be viewed as the progressive cognitive discrimination of the bioneurological algorithms from the physical level of metabolic functioning. Or perhaps, as paleopsychologist Kent Bailey, might suggest: primitive humans gradually "upshifted" from phylogenetically earlier brain centers to the control of more recently evolved higher centers of the neocortex where cognitive features not only inhibit, but also represent, rationalize, and extrapolate our more primitive urges (see Bailey 1987, 2000). Of course, as Bailey further suggests, under conditions of stress or dysfunction, humans may also "downshift" to more primitive response mechanisms. But this is not the same as sublimation.

We can now move on to the further defining qualities of the mystical experience itself: the content, the incorporative mode of the reciprocal dialectic, and whether experience is one of amplification or diminution.

CONTENT OF MYSTICAL EXPERIENCE

A mystical experience, as a subjective experience, is real. People have them and there's no point arguing about it. Something is clearly going on inside the mystic. The big problem is with interpreting the experience. People tend to describe it by whatever content they gave it in advance or by the context of the religion, philosophy, or practice through which they have it.

Therefore, a Christian mystic has a mystical incorporation with the Christ and is filled with the Holy Ghost; a Hindu or Yogi becomes one with the Brahman, or world soul; the Buddhist achieves oneness with the unitary consciousness or the Void; the Taoist grasps the unity of the Tao; the nature mystic achieves a total incorporation with nature; and so on.

Professor R. C. Zaehner of Oxford, in his *Mysticism: Sacred and Profane* (1957), which patently favors the Christian point of view, nevertheless, illustrates the many possible interpretations of what appears to be a universal experience of humankind.

Mystical experiences, as we read about them, or as we, perhaps, experience them ourselves, tend to be defined in part by the content, context, or our expectations of them. However, the common and central characteristics of such experiences are the bioneurological ones of a total extensionality and incorporation.

MYSTICISM AND THE INCORPORATIVE MODES OF THE RECIPROCAL ALGORITHM

In mysticism we are removed from the realm of pure logic and the first algorithm, in which we can passionlessly analyze and synthesize ideas and objects. In mysticism, not ideas and objects, but the ***self*** is extended and incorporated. Whenever the self is involved the reciprocal algorithm of the triune modular brain structure, the tug and pull of ego and empathy, becomes engaged. The experience of unity, then, is colored by the incorporative mode, or the balance between ego and empathy.

If the principle program engaged is ego, the experience tends to be egoistic--one of personal power with the self as the incorporating center. This experience tends to God-like feelings, or feelings of a self-incorporation of all nature. Good examples are seen in Nietzsche's will-to-power and the earlier reference to the Swami Yogananda when he feels that he has swallowed everything and transmuted it all into an ocean of his own being.

Or the personal power experience may take a more empathetically paternalistic character. An ironic, but good example of such paternalism would be our most eminent behaviorist psychologist, the late B. F. Skinner, whose hard core behaviorism recognized ***no*** inner motive states. His central character, Frasier (taken as the alter ego of Skinner himself), in the behavioristic utopia *Walden Two*, admits he likes to play God.

After showing a guest around his behaviorist utopia where everyone is manipulated and controlled into peace and happiness by Skinner's behavior technology, Frasier is asked by his admiring guest what Frasier gets out of all this. Frasier responds that he is not indifferent to power and that he does like to play God. He punctuates this, as the two stand upon a hill overlooking his paradise, by flinging his hand in a sweeping gesture that reveals his ***inner motive state*** (which behaviorism denies) and pronounces almost in a whisper

These are my children... I love them (Skinner 1962: 300).

On the other hand, if the main program engaged is empathy, the experience tends to be one of ***becoming incorporated*** with the other as the incorporating center. Examples of this experience are more common than the former in the mystical literature. The Christians become incorporated in the mystical body of Christ. St. Augustine's substance is changed into the Deity's. Ruysbroeck is eaten...lovingly and joyfully, of course. Other mystics become one with whatever they consider to be the absolute.

If the incorporative mode is that of dynamic balance of ego and empathy, the mystical union, then, becomes one of partnership with the absolute, rather than a union with overtones of domination or submission.

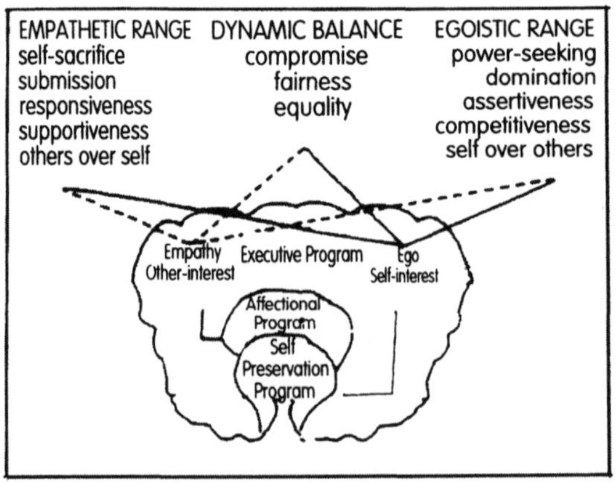

Figure10-1. The major ranges/modes of incorporative/reciprocal behavior.

DIMINUTION AND AMPLIFICATION IN MYSTICISM

Finally, a mystical experience is defined by whether we are diminishing or amplifying our activity or internal energy levels.[68]

Our brain evolved with the capacity to amplify or diminish our experience. This is a feature of protein circuitry found even in unicellular organisms without nervous systems (Bray 1995: 311). The capacity has been preserved and enhanced in our highly evolved central nervous system and is a part of everyday experience. We can amplify or diminish our feelings of power, sex, fear, anger, affection. Most often this happens rather automatically in response to events in the environment, and may be mediated by thalamocortical gating mechanisms, but we also have the power to consciously amplify and diminish willfully through concentration, meditation, visualization.

In diminution we quiet or still our metabolic energy processes by what is usually thought of as a meditational process. We put ourselves in a state of relaxed alertness, center our attention in some way, and gradually reduce our multiplicity or distractions. Reducing multiplicity is in effect also focusing and reducing physical metabolism. This kind of quiet, meditative concentration, or contemplation, is characteristic of mystical process in many religions and practices...East and West. In diminution the individual becomes completely, or nearly so, incorporated into the oneness of whatever is. All duality, or multiplicity (chaos) is eliminated in total synthesis. This is a state of tranquility and peace.

In amplification, which may, but does not have to, proceed from a state of diminution, the feelings focused upon are increased in intensity. If ego is the focus, the amplified experience becomes a subjective experience of power or God-likeness. If empathy or affection is the focus, the resulting subjective experience can be a total incorporating warmth or a becoming incorporated with an absolute spirit or god of love. Amplification, in contrast to diminution, does not reduce multiplicity but includes it all in an incorporation of totality. The Swami Yogananda's earlier reference is an example of amplification, but there are many instances in Christian, nature, and other mysticisms.

Such experiences, of course, tap limbic emotional structures, amplified by thalamocortical gating mechanisms under conscious or unconscious control. Although the disorder of temporal lobe epilepsy has given clinical insights into the possible reward or validation mechanisms of mystical or religious experience, no separate, encapsulated limbic organ of such experience has

[68]Diminution is comparable to "shutting down the senses" (Van Over 1973: xxiv). Amplification would be the reverse of this.

been identified. And, researchers are quick to note that epilepsy, as a disorder, represents a dysfunctional activation of an innate behavioral reward system that does not necessarily demean its spiritual or emotional significance. Likely, under normal circumstances, the circuitry underlying the mystical experience, as discussed previously in this chapter, is wide-ranging involving both cortical and subcortical regions with a characteristic, but as yet unspecified, limbic reward marker of a numinous mystical feeling of varying intensity (see Saver and Rabin 1997; Trimble 1996: 266-290).

PATTERNS OF MYSTICAL EXPERIENCE

The primary patterns of mystical experience clearly follow the primary algorithm. They are two.

Firstly, the meditative or diminished pattern which entails a being or returning characteristic of the I Ching, the Tao.

CHANGE

COSMIC UNITY →(increase/decrease of multiplicity caused by tension of opposites) →COSMIC UNITY

This pattern is analogous to W.T. Stace's introvertive mysticism, that we noted earlier, which emphasizes "the Unitary Consciousness: the One, the Void, pure consciousness."

Secondly, the energized or amplified pattern of becoming like that of Plato, Hegel, and the Christian mystics.

CHANGE

PRIMAL UNITY→(increase of multiplicity) →ABSOLUTE UNITY

This pattern is analogous to Stace's extrovertive mysticism, which emphasizes "the Unifying Vision--all things are one."

In sum, mysticism, in the analysis presented here, involves no supernatural variables, but only natural ones. This does not deny that there may be supernatural forces or experiences not of this world engaged. I just demonstrate that such extra-scientific phenomena are not necessary to explaining the subjective experience of mysticism.[69]

The mystical experience has all the earmarks of the subjective experience of the primary algorithm of organic synthesis combined with the self incorporation of the second, reciprocal algorithm of our triune modular brain structure, tagged perhaps with a limbic marker of ineffableness.

As a concluding note, we can see the close connection between philosophy and mysticism in the summarizing remarks of the Rumanian Marxist Dumitrui to his discussion of dialectical materialism. Dumitrui

[69]Recently a number of scholars have tried to develop a methodology for rational analysis of mysticism. See Staal (1975), Katz (1978), Katz (1992). Also a complete study of mysticism should consult the work of Rudolf Otto (1975).

states that the philosophical thought of all times can be seen as based on two rationalities. The first, like the logos of the Eleatic school of ancient Greece sees a given, permanent rationality, providing a stable basis for all change, which occurs within unchangeability. The second is like the logos of Heraclitus of ancient Ephesus which sees constant, expanding flux, opened-ended...with no fundamental permanence or stability, always self-expanding. He sees Marxism as having developed a modern dialectics in the manner of Heraclitus.

Dumitrui states that the two can be represented by two vectors with the same direction but opposite senses. The first perceives a direction toward the original, pre-existing Logos into which the whole of existence is ideally reabsorbed. We note that this has a returning quality like the Tao of China and Stace's introvertive pattern.

The second pursues a direction toward the final, to-be-created, Logos. This has a becoming quality like the more typically Western philosophical and mystical patterns. It is open-ended, however, like Heraclitus and Kant rather than closed like Plato and Hegel. This is likewise analogous to Stace's extrovertive mysticism (Dumitrui 1977, v.3: 309).

In my position, of course, each is the subjective perception or contemplation of the primary and reciprocal brain algorithms within us given different interpretations.

Again, the basic identity of the patterns is affirmed.

Chapter 11

BRAIN ALGORITHMS OF ADAPTATION, RATIONALITY, AND EXPLORATION

The whole of science is nothing more than a refinement of everyday thinking
(Albert Einstein, Nobel prize, physics, 1921)
(Einstein 1954: 260)

Following my somewhat extensive excursion into cognition, philosophy, and mysticism to trace the ubiquitous influence of the brain algorithms, it is time to return to and sum up in terms of the issues of chapter two.

Algorithms, as used in computer science and mathematics, are systematic and rigorous procedures for combining parts into wholes. Although our natural algorithms operate statistically at the molecular microlevel, they operate at high levels of probability, or rather deterministically, above the quantum, at the so-called macrolevel. At the level of conscious experience, however, the natural algorithms of our thought and behavior again appear not so rigorous. They may even seem ad hoc or careless. Why?

THE SURVIVAL VALUE OF THE HASTY GENERALIZATION

The ad hoc, careless, or hasty cognitive generalization, the creation of a whole out of parts, however, clearly had survival value in the evolutionary process. Take for example the nearly universal fear of snakes. Although the mechanisms underlying this fear reaction are not fully known, at the conscious level, they take the character of a hasty (or automatic) generalization. We immediately recoil or drawback when confronted unexpectedly by a snake. This is not a case of fine, conscious, cognitive discrimination. We don't, for instance, immediately discriminate whether the snake has a pit at its nose, identifying it as poisonous viper. We don't wait for other cues or discriminations to establish with certainty whether it is dangerous or harmless. We react first and discriminate afterwards. To wait 'til one is bitten to discriminate can be lethal.

There are other examples. An undefined shape seemingly lunging at us from the bushes...a grotesque shadow in the dark...an unexpectedly loud noise...a startling visual effect...a stranger...all provoke the hasty generalization effect. There are no fine cues, or discriminations, just gross ones. And these gross discriminations are instantaneously assembled algorithmically into a whole which we experience as danger or threat. We react to protect ourselves and perhaps discriminate later. Our brain algorithms, at some preconscious levels, are prewired, to act fast on ambiguous discriminations. Hasty, automatic generalizations may not be good science or logic, but they saved lives...especially in the primitive situations of our early existence as a species. But even today they serve to keep us unnecessarily out of dark alleys.

Neural algorithms, which are prewired to operate unconsciously and automatically, but which, reaching the conscious level, produce high level generalizations, evolved to operate in real time...survival time...when there is no time to hinder behavior by fine discriminations.[70] In less survivally urgent times, the natural discriminating, integrating primary algorithm of both our conscious and unconscious brain is manifested clearly in the development of language. Everything that can be discriminated is given its place in the order of thought and experience by a new word or name. The greater the degree of discrimination, the greater the vocabulary. And the more interesting are the wholes, concepts, theories, or syntheses, we get when we reassemble the finely discriminated parts into wholes again with

[70]Theoretical physicists Roger Penrose, a Platonist, and Stephen Hawking, a positivist and reductionist, despite their differences, confront the same issue of uncertainty in the establishment of relationships in mathematical concepts. Penrose expresses the opinion that our mathematical ability had no selective advantage in evolution and argues that it is a reasonable supposition that this ability evolved incidentally to the general ability to understand (Penrose 1994: 148-149; also 1997: 185). Hawking states that the qualities selected for in the evolution of animal brains were the abilities to escape enemies and reproduce..."not the ability to do mathematics." He further indicates the general, rather non-specific nature of the natural algorithmic processes when he adds "...It is just that the intelligence needed for survival can also be used to construct mathematical proofs. But it is a very hit and miss business. We certainly don't have a knowably sound procedure."(Hawking 1997: 177-2). This incidental evolution of sophisticated mathematical ability would be roughly supportive of what I have described in this section. The same essential argument can be seen in the discussion between the French biologist Jean-Pierre Changeux and the French mathematician Alain Connes (1995). Changeux sees mathematics as created by the brain, whereas Connes sees an archaic mathematics existing in abstract reality independent of the brain. To the former mathematics is created; to the latter it is discovered.

creativity and imagination. Often our language, our words and syntax, however, reflect hasty, automatic generalizations. They can be vague. They often lack rigor and precision.

THE GROWTH OF INTELLECTUAL AND SCIENTIFIC THOUGHT: OR MAKING RIGOROUS THE NATURAL ALGORITHM

The growth of intellectual and scientific thought is largely the history of harnessing the natural cognitive algorithmic processes and making them rigorous in the thought and behavior processes that mediate our subject/object relationship with the environment. As the review in the preceding chapters shows, the effort has been groping, and sometimes faltering.

Contemplating the primary algorithm without grasping what it was, Zeno saw change as illusory. He was stuck on the unity or ordered phase of the algorithm as the end point of the algorithmic process. Heraclitus, taking a different view, saw everything in a state of flux or change, perhaps a more accurate perception of the algorithm, but one that tended to see chaos as the end point.

Plato got more systematic about it and saw a dialectical flow process that ascended by higher and higher levels of synthesis until it included all of reality and connected with the world of ideas, which was to Plato the true reality. Plato's system of division, upon which our modern systems of taxonomy (classifications of plants and animals into ever higher groups and families) are based, followed the same model. Plato can be thought of as the inventor of the hierarchical tree later used so extensively in computer science and algorithmics.

Aristotle, only partially grasping the algorithm, tried to pin it down rigorously in his rules and procedures of the syllogism. The scholastics of the middle ages exercised the cognitive algorithm independent of physical reality and wandered off into endless and sometimes mindless abstraction.

The Renaissance philosophers, exemplified by Bacon and Descartes, saw that the cognitive dialectic alone, without being tied to reality, produced little or nothing usable in the physical world. They initiated the scientific method by adding rigor to the still undefined natural algorithmic process and requiring the results be checked and cross checked against known physical reality.

Kant dubbed the pure cognitive process, divorced from reality or experience, the transcendental dialectic. The transcendental dialectic always produced a contradiction of itself when it went off into the ozone on the quest of impossible questions. Hegel and Marx, of course, took the dialectical brain algorithm out of the head and projected it on to the cosmos and into the material world respectively.

Confusion has persisted to this day. The primary cognitive algorithm has not been properly recognized, defined, or sourced in physics, biology, and neuroscience. The second, reciprocal algorithm has never been distinguished from the first, not has it been sourced properly in evolved brain structure.

It is interesting to consider that even the present-day drive in theoretical physics toward a grand unified theory (GUT) expresses, albeit unwittingly, the primary algorithm...the effort to achieve a total ordering, incorporation, or synthesis of all nature. The algorithmic drive also comes through clearly in some, if not all, of our distinguished authors in their efforts to bring theoretical physics to the less specialized reader.

For example, in the closing page of his recent work, *Shadows of the Mind,* Roger Penrose, the distinguished mathematical physicist of Oxford University, writes in part as follows:

> ...a unity with the workings of Nature is potentially present within all of us, and is revealed in our very faculties of conscious comprehension and sensitivity, at whatever level they may be operating.

And a little further on in explaining a previous discussion of three worlds, he concludes his book with:

> No doubt there are not really three worlds but one, the true nature of which we do not even glimpse at present (Penrose 1994: 420).

Frank Tipler, a theoretical physicist at Tulane University, devotes an entire book to arguing the possibility of an ultimate unity of the intelligent cosmos in what he calls the Omega Point...a term which he borrows from the French Jesuit philosopher, Tielhard de Chardin (who is discussed in chapter 13) (Tipler 1994).

Nobel laureate Steven Weinberg, in his recent book, *Dreams of a Final Theory,* notes a motivation of physicists (to include himself), that he finds hard to explain. The motivation, which expresses the primary algorithm, seems to be a drive to turn the limited validity, tentativeness, and incompleteness of present theories into a final theory..."one that would be of unlimited validity and entirely satisfying in its completeness and consistency"(Weinberg 1992: 6).

Much earlier, Albert Einstein, who was strongly and consciously motivated by the striving for logical unity in the sciences, describes the primary algorithm without perceiving it as such. Concerning the comprehensibility to the human mind of the mysteries of the world, he writes that such comprehensibility implies the creation of

> ...some sort of ***order*** among sense impressions, this ***order*** being produced by the creation of general concepts, by relations between these concepts, and by relations between the concepts and sense experience, these ***relations*** being determined in any possible manner (all emphases are mine).

He concludes that the comprehensibility of the world is itself miraculous.[71]

The distinguished scientists and thinkers of the world experience the primary algorithm and are clearly driven by it. Else why search for order and unity? To see clearly, however, that it is an innate bioneurological algorithm, one has to step back from the drive of it and ask...why? Why am I, the human species, on this pursuit of understanding...the search for ultimate order? As the great mathematician Kurt Godel, showed us...we must step out of the algorithm itself to perceive the truth of the algorithm.

THE FIRST ALGORITHM AND THE DIGITAL COMPUTER

Rigorous treatment of the first natural algorithm reaches its height in the digital computer. Rules, procedures, instructions must be clear and precise. There can be no subjective components in the standard programmable version of the algorithm. In artificial intelligence, however, when we are trying to duplicate the natural algorithms of thought and behavior, we find that too much rigor hinders. The rules and procedures must be relaxed in rigor to produce what cognitive scientist, Douglas Hofstader, has called the "fluid concepts and creative analogies" that cognitive scientists seek to ultimately replicate in intelligent computers (Hofstader 1995).

As noted in chapter two, the hierarchical tree best represents the kind of structuring which concerns the study of algorithmics. It is a good artificial rendering of the primary algorithm, often expressed, however, with the necessary underpinning matrix left unstated. The hierarchical tree, organizes parts into wholes, in an ascending or descending (depending on preference) sequence of increasing or decreasing complexity (also as we choose).

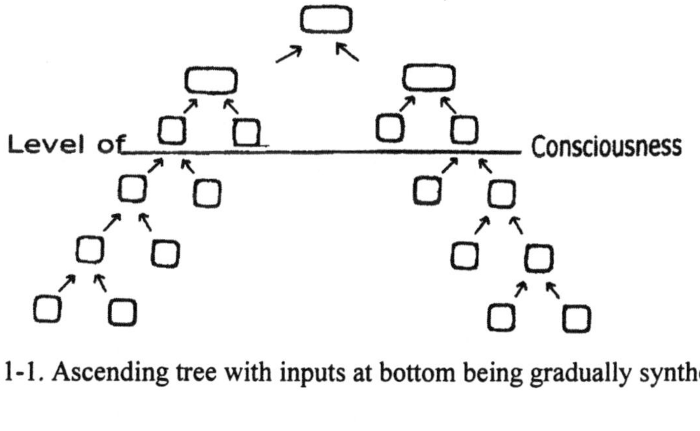

Figure 11-1. Ascending tree with inputs at bottom being gradually synthesized.

[71] Einstein (1954: 293). Hawking (1997) sees the same "hit and miss" nature of concept building as Einstein; see footnote 49.

The tree in figure 11-1 represents an ascending tree in which disparate parts are gradually assembled into the whole represented by the larger block at the top. Researchers in artificial intelligence and neural networks surmise that something like the structure depicted in figure 11-1 matches our neurological wiring.

Suppose, for this example, that multiple inputs, represented by the blocks or nodes at the bottom of the figure and also along the way up, are received from our perceptual apparatus. They are gradually integrated, ordered, or perhaps, we should say synthesized based upon their somatic linkage, into intermediate nodes or wholes. Much of this synthesis is at the microlevel and we are not conscious of the microalgorithmic process. At some point, however, the algorithmic process crosses the threshhold of consciousness. Then we are faced with a conscious process of synthesizing the micro-processed inputs into a conscious decision, action or concept.

At both the unconscious or conscious levels, however, the algorithm is the same. It is the primary algorithm of organic synthesis as replicated in our brain structure and expressed in our cognitive functioning...the ordering of chaos, the incorporation of components, the synthesizing of parts into wholes, generalizing.[72] It will be recalled from chapters 3-4, that prior to the appearance of the organic synthesizing loop, the was no such algorithm known to be present on the planet and all known life proceeds from it. And there is still no alternative algorithm.

This first, fundamental algorithm of our brain, combined with the reciprocal algorithm of the triune modular brain structure, shapes our thought, behavior, and subjective experience across the entire spectrum of human activity.

MATCH OR DECEPTION? BIOALGORITHMS AND REALITY

The specifications substrating the primary algorithm, the informational matrix of organic synthesis, in its varied expressions, probably evolved in our somatic and, ultimately, brain structure to match to a considerable degree the laws of our physical environmental reality. The perception of this innate algorithm, as framework and process, is likely the source of Plato's concept of universals, perfect ideas being the real source of their imperfect representations here on earth. The primary cognitive algorithm can be deceptive, however, in a least three ways.

Firstly, left to operate without rigor and without reference to physical reality, it can lead us on a cognitive excursion into never-never land as it did for the scholastics of the middle ages...as Kant recognized in the

[72]Hofstader (1995: 75) sees the generalizing process as an automatic, unconscious process that pervades and even defines thought.

transcendental dialectic. This was a great shortfall of philosophy and science in the millennium before the formulation of the scientific method.

Secondly, in our observations, concept-making, and theorizing we can never be sure that the order we see is the actual order of nature or physical reality or an order imposed by our cognitive algorithm itself. There is always the danger that we are deceiving ourselves, or the algorithm is deceiving us.

Thirdly, even in our disciplined thought, but especially in the largely undisciplined experiential flow of mysticism, we experience unity at both ends of the algorithm...the primal point and the end point. In the mystical experience of the Tao, for example, we experience the preexisting unity of all things. In the mystical dialectic of becoming, more common to the West, we experience the progressive and ultimate unity of all things. Actually, what has happened in both cases is that we have exhausted our capacity for discrimination. And our thought processes rest in a temporary holistic synthesis that carries the illusion of reality. Experientially, we hover in feelings and undifferentiated perceptions of oneness. For a few moments all is complete. The energized pursuit of the bioneurological algorithm is temporarily suspended...the struggle of life is over...and all is well. But soon the discriminating...the chaos...the analytical...the breaking into parts of the flowing energized algorithm begins again, bringing with it doubt, questioning, discrimination.

REALITY, SURVIVAL, AND THE DECEPTIVE ALGORITHM

We can view this natural, pervasive, inexorable primary algorithm from three perspectives.

Firstly, we may consider that we have been furnished by evolution with an algorithm that shapes everything in its own image and, at the cognitive level, keeps us from ever knowing reality as it truly is. From this viewpoint, we are either always creating our own "reality" or constantly chasing our own algorithmic tail in an ever more discriminating-integrating but circular fashion. Reality, beyond the reality of algorithm, remains forever beyond the reach of certainty for us.

Secondly, we may consider that we have been provided with a reality-matching or compatible algorithm to unlock the secrets of the universe. To discover and manipulate these secrets, though, we must harness the algorithm by rigorous procedures of logic, method, and verification. From this second viewpoint, to pursue the algorithm in physical science, we must seek greater and greater rigorous discrimination of smaller and smaller wholes into smaller and smaller parts (from atoms to quanta to strings and beyond?) and the rules for their reintegration into larger and large wholes that ultimately encompass the universe(s).[73]

Taking this viewpoint, scientists continue to pursue the search for the so-called illusive grand unification theories (GUTs) of the physical universe. To achieve such a theory, if such is possible, science still has a long way to go. But the search itself discloses the underlying brain algorithm--the quest for an ultimate and grand ordering, incorporating, or synthesis of all parts... and ***part***icles and waves.[74]

As a third alternative, we may contemplate a mix between the first and second viewpoints. We may consider that we have been furnished with an algorithm that permits us to know and experience that part of reality that we evolved to know, with the experience of the remainder being forever beyond our ken.

These three viewpoints may well persist among reasonable humans for the remainder of eternity. But, at a minimum, as we pursue this primary algorithm of our thought and behavior, we can be wary of its deceptions. We can constantly check and verify our discoveries with our best understanding of reality, so that we can understand and perhaps control that part of reality to which the algorithm permits us eventual access. Thereby we may hope to promote and perhaps establish the continued survival of our species. This was the evolutionary purpose or function of the algorithm to begin with.

A GUT-LEVEL QUESTION

The preceding chapters have touched on the second, reciprocal algorithm, but focused chiefly on the first or primary algorithm. The chapters that follow will focus more exclusively on the second brain algorithm...that of our triune modular brain structure. Before moving on, however, I might pose a final question.

Suppose our science eventually achieves the ultimate synthesis of a GUT. Will the energized primary bioneurological cognitive algorithm that has driven our curiosity, our explorations, throughout the long history of our thought and behavior, then let us finally rest in peace?

[73]How small is small? In the string theory of theoretical physics, particles are not particles at all but modes of vibration of strings. Strings themselves are estimated to be 100 billion billion times smaller than the subatomic particle called a proton(see Kaku 1994: 153).

[74]For a readable layperson's outline of the historical trek of physics combined with mathematics toward the still illusive grand unified theory of the physical universe, see Motz (1987). The author concludes that unification is still a long way off and he doubts that particle physicists are even on the right track (1987: 27). On the other hand, see the recent work of Kaku (1995) who feels that unification can be achieved through superstring theory.

Or will it still drive us, in its relentless process pattern, onward and onward... doubting and questioning... and synthesizing...and doubting? Recent developments in physics suggest that we may confront the answer sooner than later.

PART THREE

BRAIN ALGORITHMS IN ETHICS AND SOCIAL THEORY

Chapter 12

BUBER'S *I* AND *YOU*: THE INNER AND OUTER DYNAMIC OF THE RECIPROCAL ALGORITHM

On the far side of the subjective, on this side of the objective, on the narrow ridge, where I and You meet
(Martin Buber 1965: 204)

The reciprocal algorithm, the tug and pull of ego and empathy, defined by our triune brain structure is the basis as well as the dynamic of our social life as the highest mammalian life form. The energy-driven reciprocal algorithms of behavior keep us in almost constant internal conflict as ego and empathy tug and pull against each other in our daily, moment by moment, living as we interact with each other.

The dynamic which originates internally within each of us becomes externalized in our social interactions because of the effects of the behavioral tension produced in each of us and between each of us as a result of these interactions. These are the mechanisms of our social evolution (which interact with other variables when we shift academic perspectives as we have noted previously). We evolved through millions of years of social interaction with these mechanisms becoming increasingly sensitive and refined.

The outcome is that each of us, who have a fully-formed, developed, human brain, have what may be thought of as the equivalent of two persons within us...an I (ego) and a YOU (empathy). The I within us pulls us to respond first to our own needs; the YOU within us impels us to respond to the needs of others.[75]

[75]Compare Habermas (1992: 114) and Kohlberg (1984: 286). The built-in features of this programming develop effective expression in normal interactive family life.

LOVE

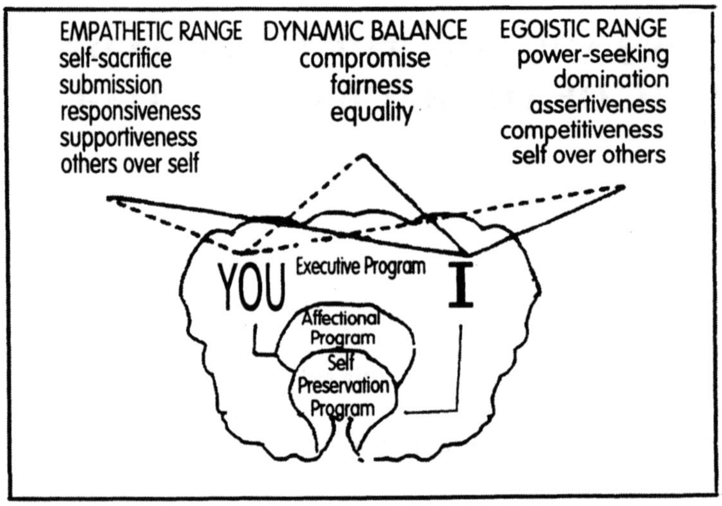

Figure 12-1. The major ranges/modes of incorporative/reciprocal behavior.

The conflict systems neurobehavioral model illustrates well this I and YOU within us. Egoistic behavior is I behavior. Empathetic behavior is YOU behavior. Wherever empathy is engaged through its roots in affection, we have the warmth of feeling, the caring, the attachment that flows from our mammalian brain structure. Where I and YOU, ego and empathy, come into dynamic balance or close thereto, we may experience both the subjective feelings and the objective expressions of what is called ***love***. We, in effect, may achieve the maxim of love your neighbor as yourself.

BUBER: I AND THOU

Of the numerous 20th century thinkers who sensed and worked to articulate this internal and external struggle, which I have defined as the second algorithm, the most perceptive was arguably Martin Buber (1878-1965), whose work had profound influence on postwar Europe.

Buber's best known work, translated into English as ***I and Thou*** was first published in 1922. The German title ***Ich und Du***, however, means simply ***I*** and ***You***. The German pronoun ***Du*** is the second person familiar form used among family members and friends. It does not carry the lofty, abstract connotation of the English pronoun *thou.*

Buber, then, came to English already somewhat misrepresented. His intent was to communicate simply, intimately. This intent got muddled, if not lost, in the translation and has tended to cause Buber to be seen primarily as a somewhat abstract mystic with the lofty thou being construed as a mystical term implying the Deity even when it referred to relations between ordinary folk.

According to Buber our interactions with the world are all driven by dialectical tension within and without. The two primary ways to interact with the world are from positions of I-It and I-You. Both hyphenated word combinations are seen as a word-pair entity. There is a tension or dynamic that binds them. Buber sees the I-It word pair as the position we take when we relate to nature, other creatures, and people as objects...objects to be utilized. On the other hand Buber sees the word combination I-You as establishing the world of relationship. One does not exploit or utilize in such relationship, but connects and experiences.

The I-You relationship occurs in three spheres or at three levels.

When with the I-You, we look to nature...animals, trees, rivers, mountains, and the like..., we do not use them as objects. Rather we connect with them in relationship

When with the I-You, we turn to other humans, who like us, share speech and concepts, we also connect with them in relationship. And we can speak to each other using the terms I and You.

When with the I-You, we approach the spiritual level, we communicate not with words, but become nonverbally aware of relationship to which we respond with thoughts and acts.

According to Buber at each of these levels or spheres of contact we approach to the fringe of the eternal You or deity or absolute. When we stand before something as I-You, whether it be tree, animal, or a spirituality, the relationship is one of love, the acknowledgment of whatever it is as a whole.

BUBER:
AND THE A PRIORI I-YOU

Buber sees the prenatal life of the child as a world-embracing union. This union is broken at birth. Throughout life, the adult retains a hidden yearning to return to this primal all-encompassing unity. Unlike the Freudian disciple, Otto Rank, however, Buber does not consider this a yearning to return to the mother's womb. Rather he sees it as a yearning toward an all-embracing union of his spiritual self with the eternal You. Striving for relation at any level, with nature, people, or deity, is held by Buber to be a priori. Buber writes about it as follows:

> In the beginning is relation as category of being, as readiness, as form that takes hold, as model for the soul. It is the a priori of relation, the inborn Thou. (Kohanski 1975: 64).

He sees the inborn You as realizing itself in contact with another, first tactile, then visual, and expressing itself in mutual tenderness. Here Buber seems to be engaging in a reification of an entity called ***inborn you*** which actualizes itself much like the ***elan vital*** of French philosopher, Henri Bergson. This inborn You is further seen as the source of the creative drive which seeks to establish things through a process of synthesis, or if failing in that, to relate to things analytically by taking them apart.

For Buber there is enormous tension and much going back and forth between I and You. His is no subtle negotiation of the razor's edge, but a dynamic and often painful stumbling back and forth from the realm of between. In his own words

> On the far side of the subjective, on this side of the objective, on the narrow ridge, where I and You meet, there is the realm of the "between" (Buber 1965: 204).

The narrow ridge as used by Buber is clearly reminiscent of the *I Ching* of Ancient China in which the yin and yang meet at the ridgepole in a state of constant tension to produce multiplicity (and chaos).

Buber's is clearly a dialectic driven by tension and since he sees love as the experience of meeting between the I and You, he is clearly describing the same process that I have dubbed the second, reciprocal algorithm. However, he sees only one dialectical process and thus confounds the primary dialectic with the second. Also he does not connect the source of the dialectical processes with physics or neuroscience and they remain a priori assumptions without a scientific foundation.

Since his grasp of the processes is only partial, the applications of his concepts are constricted in many ways. And, of course, he theologizes the algorithmic dialectic in the eternal You, giving it not only an inborn elan vital, but a cosmic one. Neither of the two can be expressed as testable empirical hypotheses.

As a final note to Buber we observe that he considered illustrative of the ideal mode of being those humans who in the passion of erotic experience are

> so enraptured by the miracle of the embrace that their knowledge of I and Thou perishes in a feeling of unity that does not and cannot exist. What the ecstatic man calls union is the enrapturing dynamic of relation...(Buber 1958: 87).

Of course, I would say it differently. The feeling of unity is the perception of the functioning of one's own bioneurological algorithmic processes of the tug and pull of ego and empathy in dynamic balance, blended momentarily in a state of amplification. A little less poetic no doubt, but tied directly to the definition drawn from brain structure and neuroscience.

Teilhard de Chardin, the French Jesuit philosopher, whom I will consider in the next section, might describe such a state as one of totalization.

FREEDOM AND I AND YOU

Before moving on, I must consider the issue of freedom, which is so vital to our modern lives.

Freedom is implicit in the first algorithmic process of protein metabolism. Organisms, to survive, must be free to incorporate the energy and materials they need. The development of extensionality, the ability to move about in search of the necessities to maintain life, which evolved so remarkably in the animal kingdom inherently carries with it the urgency to do so. Freedom below the conceptual level can be thought of as urgency as a matter of function. Anything that impedes this urgency is a matter of behavioral frustration to the organism. This frustration becomes especially apparent at the higher levels of animal evolution.

Of course, it took human beings with their big brain and its capacity for speech and thought to articulate this urgency in the concept of freedom.

Freedom is, and always will be, a concept that is a subjective product of the human brain. It may become externalized, as it certainly has, in constitutions and bills of rights, but it will always be relative and subjective. Whatever the social, economic, or political circumstances, either a person experiences herself as free or she does not.

There is another dimension to freedom, however, which is based in the triune modular brain structure...in the reciprocal algorithm. It first became implicit and potential with the appearance of the mammalian affection system. As emphasized repeatedly thus far, mammals are social animals. And humans, as the most highly evolved mammals, are the most social. Humans are born in society (even the family is a society), live in society, and reproduce in society. The mechanism which permits us to do this is the tug and pull of ego and empathy of the reciprocal algorithm.

Freedom for humankind is irrevocably and inescapably rooted in the conflicting demands, the tug and pull of ego and empathy, self and others, I and You. Because of their very brain structure, humans can never effectively escape their social life. It is wired into them by evolution. The essence of freedom for humankind, then, is in the eternal struggle for expression of ego and for expression of empathy...the struggle between the I and the You within each of us.

The I within us seeks total freedom in its expression for survival and fulfillment. But the You within us makes possible, if not demands, the acceptance of limits on that freedom.

This struggle is noticeably externalized in our political ideologies of capitalism and socialism and the post modern attempt at synthesis, democratic socialism. In a personal communication, my colleague, Paul MacLean, wrote:

> When we visited Moscow in 1986, the woman who showed us around the Kremlin ended with a little homily in which she said. "Socialism would be the best government in the world if we could only have a little freedom!"

And MacLean went on to lament

> It is too bad that society cannot combine the best of socialism and what we call capitalism (MacLean 1992b).

This lament, which echoes the sentiments of many today in all parts of the world, expresses the central dilemma of humankind...posed by the second algorithm. It is the antinomy, the eternal paradox, that we are stuck with. How well we manage it will largely determine the quality of our future.

It would be far better, perhaps, if we abandoned the isms...of capitalism and socialism, which tend to externalize and artificially polarize the internal conflict and set us apart in opposing camps. If we just recognize the issue for what it is...the inborn dilemma and challenge of the human species...we may better cooperate to make the wisest choices.

After all, we can never escape the conflict and the necessity to make choices within it. It is the secret as well as the driving force of our species and its social evolution. The unity-in-diversity with equality and freedom that we seek is our challenge to greatness which we will approach but likely never fully achieve.

Chapter 13

TOTALIZATION: THE DIALECTICAL THOUGHT OF TEILHARD DE CHARDIN

...it discovers the means of embracing them all together in a feeling which, despite its unlimited extension, preserves the warmth of human affection
(Teilhard 1969: 84)

The French-born Teilhard de Chardin (1881-1955) was both a Jesuit priest and a well-trained paleontologist. His priestly affiliation or natural inclination led him to relate the findings of evolution to Christian theology. Because of the then controversial nature of his writings, the Roman Church resisted their publication for many years. When the Church came more fully to terms with the theory of evolution, Teilhard's writing became influential. All Teilhard's major works were published posthumously.

Teilhard was a dialectical thinker who perceived evolution as a dialectical process of moving to higher and higher levels of synthesis, i.e., organization and complexity. This process was lead by what he termed the radial[76] or psychic energy of God toward an ultimate totalization and unification with Him. Teilhard saw this psychic energy as love.

[76]Tipler (1994), a theoretical physicist states that physics has identified no such thing as a radial energy. He is sharply critical of Teilhard and although he borrows the concept of the Omega Point from Teilhard, he is careful to distance himself from Teilhard scientifically. See especially (1994: 110-117).

Not only did this dialectical process work through the process of evolution from the beginning of the universe to the appearance of the human species, but it continued to operate in the behavior of humans, moving their individual and common social and spiritual evolution to a unity in diversity with the absolute.

Teilhard apparently tried to avoid the particular vitalism of Bergson's, elan vital by putting the energy of God out in front drawing us toward him, rather than working through life as a vital creative force. In this he was not totally successful. Teilhard's system still retained an aspect of vitalism.[77]

THE CONFOUNDING OF DIALECTICAL ALGORITHMS IN TEILHARD

In his thinking, Teilhard again confounded the primary and reciprocal algorithms. He perceived almost everything in terms of the second algorithm. He projected the awareness of the second algorithm as it functioned within himself onto the evolutionary process creating an abstract reification of a bioneurological phenomenon.

Since he saw love as a cosmic energy working in evolution from the beginning, he did not tie his thinking to brain structure, which appeared only much later in the evolutionary process, with its self-preserving and affectional structures as we now know them.[78]

A DIALECTIC OF BECOMING

Teilhard's dialectical concept is one of becoming. This is expressed in what he calls cosmo-genesis...the cosmos which is in the process of becoming. Becoming is also expressed in his view of the history of the universe as a cone of time. At the base of the cone is infinite, unorganized multiplicity (chaos) which converges by increased order and organization to an apex where all is united in a point which he calls Omega (and identifies with Christ) (Knight 1974: 135). Here Teilhard is projecting the primary algorithm on to the evolutionary process.

Teilhard's dialectic is reminiscent of Plato and Hegel in its ascending nature and in its becoming nature--the incorporation of all things into one great all-embracing synthesis. Similarly to Hegel, Teilhard sees the process of evolution as the transforming of matter into spirit. Hegel saw the progress

[77]See Chiari (1992: 250). Chiari notes that Bergson and Teilhard express the vitalism of the 20th century.

[78]Teilhard seems vaguely aware of the problem of a concept of love in the pre-living and non-reflective (1971: 118-120).

of history as the spirit gradually becoming actualized through the historical process (Chiari 1992: 253; also see Wolsky and Wolsky 1992).

TEILHARD AND THE CONCEPT OF TOTALIZATION

Teilhard's concept of totalization is of special interest because of its focus on individual behavior in the overall process of evolution toward unity. Teilhard sees love as the totalizing principle of human energy. The ultimate achievement of unification, he divides into three stages:

1. Totalization of individual actions
2. Totalization of the individual in regard to himself,

and finally,

3. Totalization of individuals in collective humanity.

According to Teilhard's scheme, the totalization of individual actions is made possible by love which energy both radiates from and converges in the Omega. This allows behaving individuals to see all things as having their place in the progress of life toward the Omega point. The behaving individuals who grasp this can therefore become totally attached to or incorporated with all these things (Teilhard 1969: 145-155):

> ...in such a way that they feel themselves seized and assimilated, as they act, to a far greater degree than they themselves are seizing and assimilating (1969: 148).

One could not ask for a statement more clearly expressive of the organically-based algorithm. Since he sees love as the totalizing force even in inanimate objects, Teilhard is clearly confounding the primary and second algorithms, projecting them onto the cosmos, and cloaking them in mystical theology. Stripped of its Christian theological trappings, the concept of totalization fits well into the workings of the reciprocal algorithm as I have defined it. I will borrow the concept from Teilhard, modify it, and apply it to our purposes.

TOTALIZATION: ILLUSION AND REALITY

Totalization, in our scheme of things, will be seen as the totalization of the mechanisms of ego and empathy of the reciprocal algorithm. Ego and/or empathy may be extended and incorporated infinitely to include all the cosmos, much as Piaget saw intelligence doing.

A full, complete, transcendental totalization in individual behavior can occur only, however, if ***both*** ego and empathy are fully extended and incorporate the whole of the cosmos, since anything short of a full dual extension and incorporation leaves one or the other falling short. This totalization is, of course, like that of intelligence, partly illusion and partly reality. Neither intelligence nor pure cognition, nor ego nor empathy can actually or physically be extended to incorporate the totality of reality. In that sense the experience is illusory.

But the energized algorithm of our neurological processes drives us nevertheless in that direction. And in following that drive we tend toward the subjective inclusion of all things in our intelligence, our ego and our empathy. An incorporation of all things in intelligence, ego and empathy fully extended means that we preserve and have affection for ourselves and all things equally as one.

The subjective experience of the full expression of these energy-driven algorithmic mechanisms of our brain structure is an all-embracing, extraordinarily powerful experience of love for all of creation. The subjective experience is real and it is arguably the most powerful experience available to humankind. The effects of such subjective experience may have far-reaching impact upon the world through the behavior of the experiencing individuals. From such experience comes the behavior of the great figures of humanity the world has known. And such behavior has shaped much of the human experience as we know it in history. Such a totalizing experience is what mystics have tried to describe and convey over the centuries.

In that sense, then, the experience of totalization is not illusory, but real...and very real in its effects upon the course of human history.

Chapter 14

HABERMAS: DIALECTICS, ETHICS, AND SOCIAL THEORY

For a norm to be valid...the satisfaction of the particular interests of each person affected must be such that all affected can accept them freely
(J. Habermas 1990: 20)

Discussions of ethics, whether at the individual or social levels, invariably involve the second, reciprocal algorithm of our evolved triune modular brain structure, which generally finds dialectical expression. American social theory with its positivist instrumental approach doesn't have a truly noteworthy dialectical tradition. American scholars typically gloss over or ignore the implicitly dialectical assumptions which underpin much of their positivism. In most cases we must look to continental Europe for self-consciously dialectical social thinkers of distinction.

One of the most influential contemporary philosophers from this viewpoint is Jurgen Habermas of the Frankfurt School, Frankfurt, Germany. Habermas is particularly interesting, for our purposes, because of two factors:

(1) the consistent sense of moral purpose that has pervaded his writing over the years, and

(2) his position as heir to the dialectical heritage of the Frankfurt School.

The Frankfurt School was founded in 1923 on the self-consciously dialectical tradition of Kant, Hegel, and Marx. In its early years it boasted such dialectical luminaries as Max Horkheimer, Herbert Marcuse, Erich Fromm, and Theodor Adorno. Until his retirement in the early 1990s, Habermas was the most widely known standard bearer for the Frankfurt School.

From its inception the Frankfurt School sought to develop a critical social theory free from the official dogmatic Marxism of the Soviet Union.[79] Habermas inherited this legacy and over a period of at least thirty years consistently sought to define and reformulate a critical Marxist social theory that could be applied to the changed conditions of modernity. Although the reformulated social theory of Habermas seems to bear little resemblance to traditional Marxism, it carries with it much of the same concern for moral issues and is underpinned by some of the same problematical assumptions.

STARTING FROM MARX

Marx externalized the philosophical dialectic, which is a confused and ungrounded mixture of the primary and and second brain algorithms, by moving it into society as the materialistic dialectic which he saw as expressed in class conflict. In focusing on the externalized dialectic of materialism, Marx intuitively made an assumption about the internal, unspecified dynamic of human nature...that humans would respond to social inequalities and would seek to change them once their false consciousness was revealed by Marxist ideological critique.

Marx's implicit assumption was thus the assumption of a tendency toward equality in human nature.[80] This assumption was never adequately clarified by Marx or the Marxist thinkers who followed. And it was simply ignored and buried under the weight of dogmatic Marxism with its economic determinism. This unspoken assumption was based upon the intuitive perception of the second algorithm as it tends toward dynamic balance in all of us. This hidden assumption has haunted if not guided Marxist thought since its inception.

[79]For a history of the Frankfurt School from 1923-1950, see Jay (1973).

[80]See Cory (1992) for a treatment of the assumptions about human nature underlying Marxism as contrasted with Capitalism. Marx's assumption was not only one of a tendency toward equality in human nature but also of harmony. This is seen is his view of the future communistic society of absolute social equality and social harmony. In his famous *Critique of the Gotha Program*, Marx writes the following passage on the ultimate future communist society when the individual ceases to be enslaved by the division of labor

...and all the springs of co-operative wealth flow more abundantly--only then will it be possible completely to transcend the narrow outlook of bourgeois right and only then will society be able to inscribe on its banners: From each according to his ability, to each according to his needs! (1964: 258).

In this passage Marx clearly indicates that the tendency which he assumes in human nature is a tendency toward equality and social harmony rather than one toward domination or self-interest or self-aggrandizement. Once man has freed himself of enslaving economic conditions and reached the ideal communist society, that society will maintain itself through the functioning of a principle of absolute social equality.

This same hidden assumption still underpins and drives the thought of Habermas.

BEYOND MARX AND YET?

In his early writings, Habermas, following the lead of his predecessors in the Frankfurt School, promoted two fundamental divergences from the purely materialistic dogma of official Marxism.

First, he widened the scope of social critique from a narrow focus on political economy to include mans' struggle to escape the domination of nature (scarcity) as well as the ill effects of social and cultural domination in all areas of life.

Secondly, he sought to restore the autonomy of the individual in social processes. Dialectical materialism had denied individual autonomy in favor of historical economic forces. Restoring autonomy meant backing off from the later writings of Marx as elaborated further by Engels and Lenin, returning to the earlier philosophical writings of Marx, and ultimately to a self-consciously Kantian perspective which restored autonomy, reason, and personal morality to the processes of social transformation (cf. Sartre 1976).

KANT AND THE CATEGORICAL IMPERATIVE

Kant's philosophy of morality rested ultimately on an ungrounded assertion...the categorical imperative. Kant saw the categorical imperative as the inevitable moral position that thoughtful humans came to through conscientious self-reflection upon life. It dictated that moral actions should be equally in the interest of all so that everyone should able to will such actions as universal law (Habermas 1992: 8; Kohlberg 1984:289).

Since, according to Kant, every self-reflective person would eventually and ultimately come up against the moral dictates of the categorical imperative, the imperative took on a universalistic character. Kant was, however, never able to ground it satisfactorily. Ultimately the categorical imperative was just asserted from intuition as the way things were. That's why it was dubbed "categorical" and "imperative." The resulting transcendental, ambiguous, universalism has been a major source of vulnerability of Kantian moral philosophy.

The moral imperatives of the categorical imperative, of course, derive from the dynamic reciprocal algorithm of our evolved brain structure...the tug and pull of ego and empathy as they tend toward balance. The moral imperative that Kant intuited but could not define is therefore not transcendental, but rooted in our evolved brain structure...our human nature.

THE SECOND ALGORITHM AND THE TENDENCY TOWARD EQUALITY

This tending toward balance or equality of the second algorithmic dialectic, however, as we have noted, does not inevitably lead to social equality. It can be frustrated in almost any social configuration. The behavioral tension of the tug and pull between ego and empathy can be locked in place through cultural, social, economic, and political institutions and the ideologies and rationales that support them.

To the extent, however, that ego and empathy are out of relative balance, in any social configuration, there will be domination and subordination, social and cultural inequalities of value, and the inevitable behavioral tension that is associated with them. And it is this restless tension that produces systemic frustration and drives social change when the opportunity and possibility permits. But it is more than that. It is a society-shaping force that makes itself felt, often in shifting, protean fashion at all levels of the social fabric.

HABERMAS AND KOHLBERG: IN SEARCH OF THE CATEGORICAL IMPERATIVE

Habermas's intellectual synergy with the American developmental psychologist, Lawrence Kohlberg of Harvard, is evident in the writings of both men. Following Piaget, they have both developed theories of moral development or as Kohlberg prefers "justice reasoning"(Kohlberg 1984: 224). Habermas and Kohlberg can be thought of as seeking Kant's categorical imperative...seeking a universal basis for morality or a sense of justice.

Following the lead of Piaget, both Kohlberg and Habermas posit and develop universal stages of moral development. Habermas asserts that anyone growing up in a reasonably functional family, forming his identity and maintaining himself in communicative social reciprocity will acquire moral intuitions which relate him in equity with society (Habermas 1992: 114). Kohlberg claims that based upon his empirical research the stages of moral or justice thinking, given the right social and cultural conditions, follow the same developmental sequence in all societies (Kohlberg 1984: 286). Since they are invariant from culture to culture such stages, if empirically true, must inevitably rest upon either transcendental or biological sources.[81]

[81]Habermas and his interpreters tend to be ambiguous or else dodge or gloss over the implications of their assumptions. For example see the introduction by McCarthy to Habermas (1990). According to McCarthy, Habermas is motivated to overcome moral relativism by demonstrating that "our basic moral intuitions spring from something deeper and more universal than contingent features of our tradition." These intuitions although "acquired in the process of socialization...include an 'abstract core' that is more than culture specific" (pp. ix-x). In the same work, Habermas refers to "the universal core of our moral intuitions" (p. 211). The terms 'abstract core' and 'universal core' are left undefined, in fact are admittedly intuited, and are not more enlightening than the asserted categorical imperative, which must rest ultimately either on transcendental or biological sources. In the translator's introduction to Habermas, Cronin omits entirely (1992, p. xiii) the possibility of a grounding in human nature or in biological science when he outlines the three fundamental

That is, when you say that people in every culture inevitably develop the same basic moral sense in the same stages, you are not saying anything about culture but about something more fundamental. There are only a couple of options: a transcendental cosmic source or biologically based human nature. If you choose modern science over metaphysics, you are inevitably stuck with the latter.

HABERMAS, METHODOLOGY,AND THE LINGUISTIC TURN

In his reconstruction of Marxist social theory, Habermas, however, went further than broadening its base, restoring autonomy, and seeking the categorical imperative. He took a radical methodological departure from the tradition of Marxism and the Frankfurt School.

Putting himself squarely in the Kantian tradition and yet wishing to avoid the vague universalism of Kant (Habermas 1992: 1-2), Habermas turned his methodological attention to the linguistic emphasis in philosophy initiated by Wittgenstein and the Vienna Circle. The so-called linguistic turn with its focus on the structuring effect of language and communication on human thought pervaded, if not defined, Western analytical philosophy during the mid-twentieth century.

In keeping with the new emphasis on communication which he saw as irreversible in philosophy and social science, Habermas switched from his prior Kantian ***monological dialectic*** of self-reflection to a ***dialogical dialectic*** of the Socratic/Platonic sort. He called his new dialogical methodology "discourse ethics"(Habermas 1990: 195).

The underlying behavioral assumption of Marxism however remained the same...a tendency toward equality in human nature. Habermas now sought to validate Kant's categorical imperative through the "ideal speech situation"(Habermas 1990: 196-198, Warren 1984: 175). The conditions of the ideal speech situation are a dead giveaway and link directly to the pervasive Marxist assumption. First, the principle of discourse ethics is that...only those norms may claim to be valid that could meet with the consent of all affected in their role as participants in a practical discourse (Habermas 1990: 197).

And, morality is seen as demanding equal respect and equal rights for the individual as well as empathy and concern for the well-being of one's neighbor (Habermas 1990: 200).

This definition of morality is directly expressive of the dynamic of our evolved brain structure...the reciprocal second algorithm.

---

theoretical orientations available to Habermas in reappropriating Kantian themes. So the implicit assumption of a tendency toward equality remains intuited but unexamined. Habermas (1990), however, refers to adaptation.

Discourse ethics is aimed at developing norms and morality under assumed conditions of full and complete knowledge of all effects and side effects, through the participation of all concerned freely participating, without force or domination of any kind, except the force of the better argument (Habermas 1990: 198-203).

These are ideal conditions indeed and many critics have questioned whether they could ever be achieved in any actual situation of discourse (e.g., see Cronin's introduction to Habermas 1992: xv; Warren 1984:175-176). Nevertheless, one may well ask: given the conditions of equal respect, empathy and concern for all, no domination or force of any kind, full knowledge of all effects and side effects...***would any participant choose any position of less than total equality of benefit?***

Common sense, intuition, tells us not. Our brains are so wired. And the underlying assumption of Marxism of a tendency to equality in human nature becomes inevitably confirmed by the ideal conditions of discourse ethics.

The conditions of the ideal speech situation are in fact not just neutral conditions but a set of norms. Once these conditions, or rather norms converted into conditions, are granted, the game of moral philosophy, based on the Marxist assumption, is already won or conceded.[82] All that remains is the extraordinarily difficult task of operationalizing them in discourse and the even more formidable task of operationalizing the results in any existing society.

DISCOURSE ETHICS, THE RECIPROCAL ALGORITHM AND THE PITFALLS OF A SIMPLISTIC PATH TO UTOPIA

Habermas's discourse ethics, his definition of morality, thus, are clearly based upon the intuitive perception of the second algorithm in balance. But the intuition is of a static balance, not a dynamic one. The distinction is critical. Dynamic balance is not an achieved state once and for all. It is the ever-shifting resultant of a constantly ongoing internal tug and pull, a conflict, a struggle.

Habermas's failure to source it neurobiologically and to appreciate the dynamic tug and pull between ego and empathy leads to what seems a naive oversimplification of the tasks to be done. Granting the ideal conditions is simply not enough. The dynamic of behavioral tension is a constant tug and pull even given ideal conditions.

And if the dynamic itself is not understood, efforts to achieve either ideal conditions or results may be undercut by the frustration, impatience, and

[82]This is practically acknowledged by Apel (1988: 348) when he writes in part that anyone entering into augumentative discourse on questions of morality has, by the act of entering, acknowledged "the rules of cooperative argumentation and hence also the ethical norms of a communication community". This is quoted by Habermas himself (1992: 76).

ultimate disillusionment produced by faulty understanding and unrealistic expectations. Such frustration, impatience, and disillusionment can lead, as it has before, to the imposition of ideology, ever well-intended; and with the imposition of ideology, to eventual oppression and totalitarianism.

The path toward Utopia is not just a matter of eliminating false consciousness through self-reflection or communicative dialogue and then making it happen.[83] We must forever struggle to manage the restless dynamics of our evolved triune brain structure and the effects of those dynamics as they become externalized under conditions of scarcity in our social institutions.

Since the dynamic algorithm is everpresent and tends to equality, it is seductive to view the fulfillment of the tendency as inevitable. It is not. We will approach the ideal only by a full understanding of the natural dynamic, how it externalizes in society, and by making the conscious choice to guide our social evolution in that direction.

SUMMARY

In sum, here is what Habermas did to the Marxist dialectic. He abandoned dialectical materialism for a return to a dialectical Kantian emphasis on individual autonomy and reason. Then he abandoned the self-reflective dialectic of Kant for an externalized communicative dialectic of the Socratic type. But it wasn't a simple going back to the ancient form. He took with him the implicit Marxist assumption of a tendency to equality in human nature which added a new element...an element that had never existed in the methodology and thought of Socrates and Plato.

[83]On the utopian potential of modernity, see Habermas (1987). Also see Seidman's introduction to Habermas (1989).

[illegible]

ultimate likelihood [illegible]

[illegible]

social institutions.

[illegible]

[illegible]

PART FOUR

BRAIN ALGORITHMS IN PHILOSOPHY AND LANGUAGE

Chapter 15

THE PRIMARY ALGORITHM IN CONTEMPORARY PHILOSOPHY

We can cheerfully admit that any such model-building must grant that nature has wired in some unacquired abilities to perform higher-order mental operations.
(Richard Rorty, Post analytic philosopher, 1979: 241)

It is appropriate to begin this section on philosophy, language, and cognitive neuroscience with a look at the expression of the brain algorithms in contemporary philosophy.

Contemporary philosophy, beginning with the so-called linguistic turn in the 1950s, had, for the better part of three decades, been greatly concerned with questions of language analysis. After, in the judgment of many, having exhausted its language programs, it has largely emerged from linguistic analysis to define new programs for itself. During the linguistic period it dealt with, and in some quarters still continues to deal with, many questions of language which are now being actively taken over by the formal study of linguistics and cognitive neuroscience.

Contemporary post analytic (linguistic) philosophy, like the more traditional logic and dialectical systems of thought, illustrates clearly the guiding, dynamic framework of the primary bioneurological algorithm. Although a comprehensive review of the new post analytic philosophy would be well beyond the scope of this chapter, some observations can be made looking at the work of one of its leading scholars, humanist and philosopher, Richard Rorty.

RORTY, DECONSTRUCTION, AND THE MIRROR OF NATURE

In his *Philosophy and the Mirror of Nature* (1979), Richard Rorty, one of the leaders in the so-called deconstruction of the analytic school that even yet continues to be predominant in Western philosophy, argues against the project of analytic philosophy and traditional concepts. Among these concepts are the mind-body dualism of Descartes, the image of the mind as a mirror of nature, the philosophical foundationalism of Kant, and the Platonic concept of mind as the organization of universals.

Rorty argues for what is essentially a philosophy of flow and change, a philosophy which he opposes to systematic philosophy. Rorty refers to this philosophy of flow and change approvingly by the term ***edifying***. The use of the complimentary term apparently reflects Rorty's approval of his preferred flowing outlook in contrast to the systematizing philosophy, of which he disapproves, and which he claims tends to shut down the flow.[84] Rorty seeks to keep what he calls the ***conversation of mankind*** going by varying and critiquing the conversation from multiple perspectives (poetry, literature, art, etc.), rather than building systems that seek ultimate truth in the form of universals.

The features of flow (change), systematizing (synthesis), that permeate Rorty's discussion are clearly characteristic of the primary algorithm. Although he makes mention of issues of morality and ethics, his thought in this work scarcely touches the second, reciprocal algorithm of our triune modular brain structure. Despite the fact that he does not recognize or define the first algorithm anywhere in his writing, the algorithm, nevertheless, clearly operates as the dynamic framework of Rorty's thought.

In opposing systematic (or we might say synthesizing) philosophy, Rorty is opposing the blocking of the flow of the algorithm, toward any final and ultimate syntheses or concepts (or incorporations). He is subscribing to Charles Sanders Peirce's well-known admonition that we should not block the path of inquiry (see Hartshorne, et al. 1933-58, v.1: 135). He rightfully sees the aim at achievement, or identity, or discovery of such synthetic universals leading to the danger of ideology or limiting the creative development of humanity.

Ultimate or final systems do carry the threat of ideology. They discourage questioning and, in the employ of true believers or cynical or

[84]Rorty (1979), describes Wittgenstein, Heidegger, and Dewey as having the aim to "edify"; that is, "to help their readers, or society as a whole break free from outworn vocabularies and attitudes, rather than to provide 'grounding' for the intuitions and customs of the present." Philosophers who attempt to do grounding fall into Rorty's class of systematizing philosophers, whose motives to help readers and society he seems to implicitly impugn (1979: 11-12, 386-387).

even well-intended bureaucrats or autocrats, can lead to tyranny...just as religion and other forms of ideology (e.g., communism, fascism) can. To stop, to repress, block, or dam up the algorithmic process leads to stagnation in philosophy, rigidity in the arts, tyranny in politics.

From this viewpoint of the necessity of movement, a flowing conversation, Rorty's implicit dialectic is in the nature and possibly the tradition of Heraclitus. The moving or flowing conversation is like the metaphor of life as a river...a metaphor used by Heraclitus and characteristic of the Tao. Rorty's implicit dialectic would be more compatible with the former, however, than the latter. Heraclitus's dialectic flows endlessly, always open, never achieving a full and final synthesis, always changing through opposition, strife, or possibly just further discrimination or doubting. Rorty emphasizes the movement of the algorithm, the chaotic aspect, and he does not want it to stop. The Tao, on the other hand, although sometimes described as a river, sees change as somewhat illusory in the manner of Zeno, always occurring within the context of an underlying unity.

Rorty argues against the image of the mind as a mirror of nature. He takes a position that he calls epistemological behaviorism that eliminates any dualistic or mentalistic concepts. He does not, however, see the brain as a tabula rasa, or blank slate. In his chapter "Epistemology and Psychology," he writes that in building models of behavior

> We can cheerfully admit that any such model-building must grant that nature has wired in some unacquired abilities to perform higher-order mental operations. At least some of those little men performing subroutines in various brain centers will have to have been there since birth. But why not? If one gives up the notion that empirical psychology is going to do what the British Empiricists failed to do--show how a tabula rasa gets changed into a complicated information-processing device by impacts upon peripheral sense organs--then one would not be surprised that half of the adult's subroutines were wired into the infant's brain on instructions from the chromosomes....(Rorty 1979: 241).

The foregoing statement could be taken as anticipating or referring to the bioneurological or brain algorithms. Once we understand the preprogrammed algorithmic structures and functioning of the brain, we can, of course, agree with Rorty, that the metaphor of the mind or brain as a mirror of nature is wrong. The algorithms do not reflect, but rather, shape our perceptions of nature into usable forms for survival of the human organism they serve. Photons or light waves of different wave lengths, for instance, are discriminated and interpreted as colors by algorithms of our visual neural apparatus. The survival purpose of such colors is to allow us discrimination of aspects of our environment so that we may identify and respond to them appropriately. If the brain is anything, it is a discriminator and an organizer...an analyzer and a synthesizer. The brain discriminates and organizes the sensory data from the environment and gives us, not a mirror reflection of the environment, but an algorithmic interpretation of it.

In algorithmic terms, it seems Rorty is saying that systematic philosophy aims at ultimate synthesis, whereas edifying philosophy aims at extension, chaos, change. The two aims, taken separately, split the algorithmic process. You can't have one without the other. By focusing on one or the other, you create distortion.

The term "deconstructing" used by Rorty (and others) to describe their process of dismantling traditional philosophy really just means "taking apart"--one of the two fundamental alternatives of the algorithmic process. The use of the former term is rather faddish. It gives the misleading impression that one is doing something new. Actually, taking apart an old synthesis is what thinkers have done repeatedly throughout history. Generally, they take apart to rediscriminate and reassemble (by proposing new or different relationships among the parts) into a "newer" synthesis of a somewhat similar nature.

Rorty claims not to be putting the parts back together into a system or whole. But this seems to be self-deceptive.[85] He may not be building what he calls a system, but he is certainly synthesizing. He just doesn't emphasize synthesis, indeed denies it, and therefore distorts his perspective. His reconstructed system has what he refers to as a holistic perspective. This holistic (synthetic) perspective joins all inputs (parts) into a system of flow, of ongoing conversation.[86] The larger whole that Rorty posits is the conversation of mankind with its many voices (parts) included in the global holistic dialogue. It all follows the inescapable framework/dynamic of the primary algorithm.

PLENTY OF ALGORITHMIC ROOM FOR FLOW, FOR DOUBT

In all aspects of the primary algorithm there is room for doubt. Any discrimination of particulars (parts), description of relationships, and resulting syntheses are vulnerable. In the particulars, or parts, we can doubt the authenticity or reality of the objects or parts themselves. After all they are the algorithmic discriminations of our perceptual neural apparatus and

[85]As Moser notes, philosophers who shun explanation end up, contrary to their claims, offering explanations (syntheses) (1993: viii, 3).

[86]Rorty (1979: see 174, 181). Rajchman (1985: xi-xii) points out that Rorty's holistic pragmatism stresses Dewey's perspective that there are many communities of inquirers rather than Peirce's single logic of inquiry. In a later work, Rorty calls his pragmatism a philosophy of solidarity not objectivity (1985: 1-19).

there is plenty of room for variety and some arbitrariness in that. In the relationships we establish between and among the parts there is, again, plenty of room for arbitrariness, doubt and disagreement. Perhaps the margin for ambiguity is greatest in our describing of the relationships...after all, Einstein admitted that we do it in any way possible (Einstein 1954: 293).

The syntheses that are created from the dubious parts and their dubious relationships are thereby inherently doubly dubious. This is especially true when the discriminations and conceptual entities (syntheses) are those of philosophical speculation which reaches beyond what can meet reasonable scientific standards of verification.

We then fall into the situation of the transcendental dialectic in dealing with such impossible or inherently dubious questions. As Kant so well knew and modern philosophers so easily forget, pure reason always proposes an opposite position which is equally acceptable to pure reason. Such is the nature of the algorithm. As noted previously in chapter 11, when we play the transcendental game, i.e., deal with questions not subject to empirical verification, we are at the mercy of the unanchored algorithm. We are either always re-creating our own tentative new reality or chasing our own algorithmic tail. The former phrasing lends more dignity to the philosophic enterprise, but the latter probably more accurately describes much of the history of philosophical inquiry.

The eternal recurrence...despite spectacular and sometimes fruitful, side trips and diversions... philosophical thought always returns to fundamentally unanswerable questions...unanswerable at least at the level of philosophical speculation. Once they become answerable, they tend to move into the province of science. Or as linguist Noam Chomsky puts it, concerning questions of the mind, in terms expressing unwittingly the primary metabolic algorithm:

> ...the concept of "physical explanation" will no doubt be ***extended*** (emphasis mine*)* to ***incorporate*** (emphasis mine) whatever is discovered in this domain, exactly as it was extended to accommodate gravitational and electromagnetic force, massless particles.... (Chomsky 1972: 98).

It is interesting in the above passage to note the primary algorithmic terminology of extension and incorporation used by Chomsky although he has no concept of such an algorithm.

THE PLATONIST, KANTIAN,AND POSITIVIST ILLUSION

At the opening of the final chapter, *Philosophy Without Mirrors*, Rorty sees it hard to imagine what philosophy would be like without the traditional concepts which he criticizes and advocates abandoning. His remarks are very revealing of the primary algorithm itself. He writes:

> ...The difficulty stems from a notion shared by Platonists, Kantians, and positivists: that man has an essence--namely to discover ***essences*** (emphasis mine). The notion that our chief task is to mirror accurately, in our own

> Glassy Essence, the universe around us is the complement of the notion, common to Democritus and Descartes, that the universe is made up of very simple, clearly and distinctly knowable things, knowledge of whose essences provides the master-vocabulary which permits commensuration of all discourses (Rorty 1979: 357).

And Rorty goes on to say that this "classic picture of human beings must be set aside before epistemologically centered philosophy can be set aside"...to make way for hermeneutics or edifying philosophy.

What Rorty has described unwittingly is the illusion-generating quality of the primary algorithm. The fundamental discriminating-ordering, extending and incorporating, function of the algorithm drives us almost inevitably in the direction of such notions quoted above..."that man has an essence--namely to discover essences." The algorithm creates order from chaos, wholes from parts, and vice versa. Small wonder that Platonists, Kantians, and even positivists, riding the algorithm unconsciously, got caught up in the illusion...the illusion of "essences"...of discrimination and ordering even when there is nothing to be discriminated or ordered except the algorithmic function itself. As I have pointed out from the beginning, it is characteristic of the brain's primary algorithm that it seeks to find order, and failing that, imposes order (cf. Gazzaniga 1998: xiii, 156-159).

The search for foundations of knowledge, for essences, for total incorporations, for total syntheses, for unity, manifests the algorithmic drive. The drive, the search is the clearer reality. We encounter illusion when we take too seriously, hold too firmly, to the particular discriminations and then to the tentative incorporations, concepts, syntheses, unities, universals that we temporarily formulate along the path of the search. Such is the nature of the primary bioneurological algorithm. The second algorithm is different. As the social dynamic of our triune modular brain structure, it has the materiality, the structure and dynamic upon which we build our social and moral lives.

OF RESPONSIBILITY FOR THOUGHT
THE VALUE LINKAGE

So Rorty has given up and seeks now only to keep the conversation, the algorithmic process, flowing. He states that all we need is the method of cultural anthropology (Rorty 1979: see the discussion on 238). In another context Hilary Putnam has observed that the giving up of traditional standards raises the question of whether a path can be found between "the swamps of metaphysics and the quicksands of cultural relativism..."(Putnam 1983: 226), which relativism is the hallmark of cultural anthropology.

But Rorty is still a captive of the algorithmic process even in his new position. To use any and all input from any and all actors and sources, poured indiscriminately into the ongoing conversation of humankind, without attempting to construct (synthesize or systematize) anything from it, is to effectively avoid the responsibility for thought. In all fairness, I am not sure this is what Rorty intends. But I wholeheartedly agree that we must not dam up the algorithmic flow process, once we understand what it is. The price to be paid in the build up of behavioral tension, the stultification of creativity, and the oppressiveness of any relevant political aspects, would be too high and probably untenable over the long haul.

We can, however, take responsibility for our thought. We can flow, change, create and recreate ever new syntheses, recognizing in the nature of the algorithm, such syntheses will always be held tentatively, never dogmatically, and constantly reexamined in the flow of conversation. We should synthesize with care, knowing the vulnerabilities. And these primary algorithmic thought processes must always be subject to check by the dynamic elements of our triune brain structure...the tug and pull of our programming that seeks the survival and prospering of self, our species, and the world of which we are a part and upon which we depend for our existence.

It is the inevitable linkage of the second, reciprocal algorithmic matrix of our triune brain structure to the primary algorithm that demands we be responsible. The questioning, the doubting, the synthesizing is crucial to our species' survival, and at bottom, despite some superficiality, bears on issues of our ultimate survival...the informational DNA matrix is inevitably there. The second algorithm of our triune brain structure (the expression of the mechanisms of self-survival and affection or survival of the species) constitutes the essential protein expression of the genetic matrix and assures the link between fact and value that must always obtain.

And it does apparently obtain implicitly with Rorty. Rorty is apparently falling back upon what I have defined as the reciprocal algorithm for justification of pragmatism when he claims that his pragmatism wishes to increase the "good" for "us" to "better" for "us" and to extend the reference of "us" as far as we can (Rorty 1985: 5). We may look to chapter 13 herein on Teilhard and the concept of totalization for a comparable transcendental version of this same motive. Of course, as a pragmatist Rorty gives "good" and "better" no definitions and so they are completely relative as he uses them.

Rorty is also invoking the second algorithm when he says that as a partisan of solidarity, the pragmatist's (meaning Rorty's) account of the value of cooperative human inquiry has only an ethical base, rather than an epistemological or other (Rorty 1985: 11,16). Rorty, trained in the linguistic, analytical tradition (which he now rejects), and Habermas (see chapter 14), converting to it in his later work, both want ***good or better for us*** to emerge out of a conversation of all voices. Unlike Habermas, however, Rorty has no methodological counterpart to the former's discourse ethics. Rather, Rorty expects to get from good to better for us by just "muddling through"(Rorty 1985: 11). But the very fact that he thinks we can get from good to better in this muddling way suggests that he is yet carrying an implicit and unacknowledged Kantian notion in the back of his head...that of the categorical imperative... an imperative that all ***voices*** of humankind will come up against in the process of open and free conversation much as Kant assumed all ***reasoning*** men would come against it in the process of reasoning. Without such an implicit notion, there is no reason to assume that the conversation would not equally lead us from good to bad to worse. Only the vocabulary and perspective have changed...from the internal and private monological reasoning of Kant to the social conversation or dialogue of the linguistic analytic emphasis.

In closing this chapter, I may reiterate that my purpose has not been to critique analytic or post-analytic philosophy. Nor has it been to trace or critique the development of Rorty's thought. The purpose has just been to demonstrate the pervasiveness of the primary algorithm in contemporary philosophical analysis.

Chapter 16

BRAIN ALGORITHMS AND HUMAN LANGUAGE

...The raw material [of language] *is of no use unless it can be broken down as food proteins are broken down into amino acids, and built up again into the pattern of... indwelling latent structure*
(Eric Lenneberg, M. D. 1967: 378)

The two bioneurological algorithms of protein-nucleic acid synthesis and our triune modular brain structure, underpin, and shape the structure, process, and meaning of language in all human societies. They play differing but complementary roles. As in other aspects of our thought and behavior, the primary algorithm provides the framework for and is pervasively expressed in all language. As the primary cognitive process/framework the primary algorithm structures the relationship between language and thought. The second algorithm, on the other hand, as the essential somatic matrix shapes the semantic significance or value component of language. Each of the above claims will be examined in detail.

SYNTAX ANDTHE PRIMARY ALGORITHM

Syntax is defined in linguistics and ordinary grammar as the structure of sentences. The structure includes the basic framework as well as the rules for putting sentences together. The basic syntax (structure) of language and patterns of thought are the same.[87] They both follow the framework of the

[87] For a recent interdisciplinary consideration of the relationship between language and thought, see Carruthers and Boucher (1998). The probable relationship between logic and language has been much discussed. Swiss psychologist, Jean Piaget, whose work was discussed in chapter 7, held that the standard logic, innately based and expressed developmentally, made up the formal laws of thought (Piaget 1953, 1970; Inhelder and Piaget 1958).

primary algorithm. That is: first, we discriminate (excerpt from the whole), then we relate, then we reassemble or synthesize. Then we describe in language...assigning the phonetics (sounds or utterances) or written conventions that have become accepted or agreed upon in any language community.

It is significant that the grammatical definition of a sentence is that it expresses a complete thought (synthesis). The very definition indicates the intuitive grasp of the algorithmic unity of thought or cognition and language. All sentences are syntheses...groups of words (parts) held together by the glue of relationships into a whole. As a whole synthesis they make sense to our cognitive structure. In this they demonstrate the synthetic aspect of the primary algorithm.

Incomplete sentences, on the other hand, demonstrate the chaotic aspect of the primary algorithm. They don't make a whole. They don't make sense to us. They leave us hanging...in the chaos or extension phase of the algorithm. Reflecting the energized nature of the algorithm, incomplete syntheses or sentences leave us in a state of tension. The indeterminacy of chaos is uncomfortable, frustrating...although it may also be experienced as stimulating or challenging at moderate levels. Chaos, of course, means that we lack either the parts or relationships among the parts to make a comprehensible whole, which would be a complete sentence...a complete thought.

The fundamental pattern of syntax likewise follows, then, the primary algorithm

subject
thought→ / (verb) →sentence (synthesis)
object

incorporation→ extension → incorporation

thesis
synthesis→ / (relationship) →synthesis
antithesis

order → chaos → order

As unjoined parts, the grammatical subjects, objects, of any sentence reflect the previous orderings of previous discriminations. The verb

---

Others reject the notion of such a formal system (e.g., Cheng and Holyoak 1985; Johnson-Laird 1983). More specifically, similarities between logic and syntax itself have often been noted, and the as yet unclarified relations between syntactic structures and logical structures have been discussed extensively (e.g., Falmagne 1990, Hornstein 1984, Sommers 1982).

establishes the relationship, the action, the state of being that exists between the grammatical subject and the object(s), which binds them into the synthesis...of a complete sentence. Given the assertion of just a subject or object in isolation, the utterance or thought is in a state of chaos or extension, until a verb provides the relationship and/or attaches it to another object.

THE PARTS OF SPEECH: THEIR PRIMARY ALGORITHMIC STRUCTURE

In English speaking countries, the standard grammatical parts of speech that we all learn in elementary school reflect the primary algorithmic pattern of subject---verb---object combined with the determiners and qualifiers that cluster about them.

1)Nouns--tag or name discriminated or synthesized objects or subjects, concrete or abstract. Nouns may be tagged at any level of generalization or synthesis...from a small, perhaps basic particle like quark up to an all-bracing synthesis like universe, which includes innumerable billions of quarks as constituent parts. Pronouns are, of course, substitutes for nouns.

2) Verbs--describe relationships, change (chaos), movement

3) Determiners and qualifiers

a) adjectives--qualify nouns (objects)

b) adverbs--qualify verbs (changes, relationships)

c) articles--designate nouns

d) prepositions--designate object relationships or change

4) Conjunctions—connect; establish relationships of hierarchy or equality

When we make a sentence (language synthesis), we take one or more objects and establish or express a set of relationships among them. The determiners and qualifiers are bound to their particular subject, verb, or object and must cluster with them or be tied to them by some form of reference. British linguist, Geoffrey Horrocks, in his book on generative grammar, indicates unwittingly the primary algorithmic basis of syntactic rules when he writes:

> ...In other words the rules express part-whole relationships, and impose an order on the parts. This information is often represented in the form of a tree diagram (Horrocks 1987: 32)

As indicated in chapters 2 and 11 of the present text the hierarchical tree represents the kind of structuring which concerns the study of algorithmics. It is a good artificial rendering (with matrix often implicit) of the primary bioneurological algorithm. In fact a review of texts and monographs on grammatical theory shows that the hierarchical tree is, overwhelmingly, the preferred way to illustrate the parts of a sentence or phrase and relationships that bind them into a whole.

PARSING AND GENERATING: THE ALGORITHMIC NATURE OF GRAMMAR

Throughout the chapters of this book, the point has repeatedly been made that we have no other way to think outside of the framework of the primary algorithm. In successive chapters, the pervasive algorithm has been disclosed to lie beneath the alternative vocabulary of each separate discipline. In others words, we are saying the same thing differently, in the numerous vocabularies of the various disciplines. The case is also true of the study of theoretical linguistics. In that discipline, also, we have only the primary choices of taking things apart or putting them together. Nowhere is this fact more clearly stated than by Lauri Karttunen of the Artificial Intelligence Center of SRI International and Arnold Zwicky of the linguistics department, Ohio State University.

In discussing new notions of language parsing, Karttunen and Zwicky state that with the movement from traditional to more formalized grammatical descriptions parsing of sentence structure could be seen as algorithmic. The basic rule or principle of sentence structure expressed as

S (sentence) = NP (noun phrase) + VP (verb phrase)

or as a binary tree,

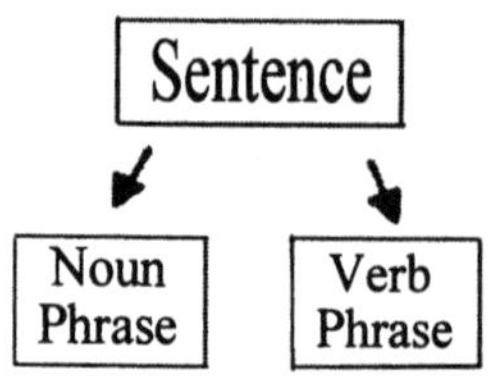

They explain that the full set of principles expressing the foregoing relationship

> ...constitutes a *formal grammar* for the language, which can be seen, indifferently, as having an analytic or *parsing function*, or a synthetic or *generative function* (italics in original) (Karttunen and Zwicky 1985: 3).

In other words in linguistics, when we ***analyze***, or take things apart, we call it ***parsing.*** And when we ***synthesize*** or put things together, we call it ***generating***. The algorithm is the same; only the words are different.

ALGORITHMIC INTEGRITY AND THE FLEXIBILITY OF SYNTAX

Other things being equal (e.g., semantics and context), the grammaticality and intelligibility of a sentence may be said to rely upon what I choose to call its ***algorithmic integrity***...the integrity of the parts and their relationships. As long as the algorithmic integrity of the S = NP + VP, or more traditionally, the SVO relationship (to include any of its variants of

SOV, VOS, VSO, OVS, or OSV,[88] depending on the specific language being examined) with its associated clusters of qualifiers and designators, is maintained, the sentence is both well-formed and, if semantic and contextual conditions are met, intelligible.

Furthermore, provided algorithmic integrity is maintained, the various parts of the sentence may be moved about to any position in spoken or even written expression and they will be intelligible. Algorithmic integrity is broken when the elements of the basic triadic framework cannot be clearly distinguished. If the algorithmic integrity of the structure is broken, the utterance, verbal or written, becomes neither well-formed nor intelligible.

Why is this so? Because our cognitive structure can only process data on its own algorithmic basis. Of course, in keeping with the cognitive tendency to impose order when it finds none, our cognitive algorithm may attempt to impose its own order upon the broken or incomplete utterances, but the intelligibility will remain dubious...guesswork.

Renowned linguist, Noam Chomsky, has long operated from the hypothesis of an innate pattern or structure for grammar. Almost four decades of research under the Chomskyan program, however, although producing numerous insights, has not adequately established the nature of this innate grammatical structure.[89] Other theories, such as *Generalized Phrase Structure Grammar* (GPSG) (see Gazdar, et al. 1985), have denied the psychological reality and the helpfulness of the concept of a universal innate grammar. I hold with Chomsky that there is an innate grammar. And further, that the pattern of the innate grammar is the same pattern as that of the primary algorithm.[90] My contention here is that the only necessary constraints are those needed to maintain primary algorithmic integrity.

[88]Or even SV or subject-verb alone, like *she swims*, or, *he gives*. Each case can be a complete sentence with the direct and/or indirect objects implied. After all, she has to swim somewhere and he has to give up or to something. At the very least both sentences imply an action that takes place in an environmental context. In linguistic theory sentences are viewed in a somewhat different terminology than the traditional grammatical categories of subject, predicate (verb), object. A sentence is seen rather neutrally as composed of a noun phrase (NP), which can be a single noun (usually designated as subject traditionally) or a noun with all its determiners and qualifiers (articles, adjectives, et al.) plus a verb phrase (VP), which may be a single verb or a verb with its determiners and qualifiers (adverbs, prepositions, et al.) as well as its direct and indirect objects. In all the above cases, however, even where context is implied, if primary algorithmic integrity is maintained, the sentence is intelligible.

[89]See Chomsky's numerous works, beginning with his early work on syntactic structures (1957) through his later works (1986a and 1986b). The bibliography lists these and some other key works.

[90]Compare the discussion in Pinker (1994: 41), before it goes further than the primary algorithmic essentials.

Philip Lieberman, writing from an evolutionary biological perspective, supports the concept of innate mechanisms that structure and facilitate the acquisition of language. He criticizes, however, the rather complex, tightly interlocked system of Chomsky as biologically implausible.[91]

In his discussion of learning to talk and think, Lieberman argues that although specialized brain circuits clearly underlie speech production and syntax, many, if not all, the rules of syntax and motor control patterns are learned by means of general cognitive mechanisms. Among such mechanisms are associative learning, imitation, and hierarchical categorization (Lieberman 1991: 134-146).

The primary incorporative algorithm, as replicated in cognitive structure, is the foundation of associative (incorporative) learning so familiar to students of animal and human learning. Imitative learning, often seen as a separate mechanism from associative learning, is also incorporative in nature and follows the primary algorithmic framework and process. And, of course, as pointed out in chapters 2 and 11 especially, hierarchical categorization is part and parcel of the primary algorithmic structure. More will be said about this in the discussion that follows.

Martin Braine, of New York University, has developed what he calls a "natural logic" approach to reasoning which he sees as an innate cognitive faculty that shapes the development of language (1990, 1992, 1993, 1994). Braine's schema maps on the primary algorithm rather well. His natural logic, based upon studies of language development in children, requires essentially just nouns and verbs, with objects (nouns) related to each other by the verbs. Braine uses the algorithmic hierarchical tree to illustrate his concept (1994: esp. 26-27). Braine builds his position on the assimilation/ accommodation concept of Piaget, discussed in chapter 7. This further substantiates it as an incorporative, primary algorithmic scheme.[92]

SELF REFERENCE: WE ARE ALWAYS THE SUBJECT OF OUR SENTENCES

In keeping with our subject/object relationship with the environment, prescribed by the primary algorithm of the organic synthesizing loop and as replicated in cognitive function, the pattern is always, invariably, self referential. That is, in our discriminations of the environment we always relate them to ourselves. This is in keeping with the necessary, everpresent somatic informational matrix. We are always the subject of any relationship.

[91]Lieberman (1991: 130). For argument against the mental representationalism implicit in positing a universal grammar, see Harris (1987: 131-132).

[92]For a series or articles which proceed from or build upon the incorporative scheme of Piaget, see Overton (1990). See Falmagne (1990) for a position between Piaget and Braine which sees logical knowledge as deriving from linguistic as well as nonlinguistic sources.

We are, therefore, always the subject of our sentences. This is assured by the subject/object relationship of the primary algorithm. We can pretend detachment or objectivity by obscuring or denying our connection, by an artificial "objective" externalization, but the relation is always there. Because it is we who are creating the discriminations, the syntheses, the language sentences. The grammatical subject as we learn it in formal language instruction is an artifact or convention of written language.

For instance take the sentence: *The cat chased the rat.* The grammatical convention of written language names the cat as subject. The complete sentence or thought, however, would be *I observe*, discriminate, etc., *that the cat chased the rat.* ***I*** is the true subject and it is I who discriminated that the cat (one object) chased the rat (another object). I am the true subject of all sentences that I create. I communicate with you by making you a second subject of my sentence. That is: I communicate to you that I observed that the cat chased the rat. These are somewhat trivial examples but illustrate our inevitable subject/object relationship with the environment.

The ***primary algorithm***, then, establishes the basic ***syntax*** of language and the basic subject/object, self-referential nature of all sentences. But there is more to language than just syntax and self reference. For an understanding of meaning and value or significance in language, we must go the three dimensional protein expression of our evolved genetic informational matrix...the second, reciprocal algorithm. Such elements of language study fall generally under the rubric of semantics.

THE RECIPROCAL ALGORITHM AND THE NATURE OF VALUE

All thought and language springs from our subject/object relationship with the environment and the externalizations of it; that is, from our primary algorithmic interactive relationship. We are each the subject of all our sentences and the sentences are only significant in that relationship.

The second, reciprocal algorithm provides value to words and to language. All basic words are learned and coded to the master or subroutines of the self-preservation and affectional programs of the second algorithm. In noncomputer science terms they are coded to or describe our basic biological needs, ultimately prescribed by and irrevocably tied to the genetic information matrix.

For example, as an infant I perceptually discriminate, by sight, sound, touch, taste and smell, an object that moves, cares for me, nurses me, warms me, comforts me, and I learn to relate the sounds, ***ma-ma*** (in English, anyway), to the object. The word is coded to my several basic needs, across the spectrum of many subroutines, and bears all the meaning of the elements of self-preservation and affection. It codes to what will become my highly generalized ego and empathy. It is a powerful word, loaded with meaning, significance, and value.[93]

Take another example. I, later in my yet short life, see (visually discriminate) an object in the environment with me. I learn that the object of my discrimination is called ***dog***. The dog has value to me fundamentally as a part of my environment. And within that fundamental relationship, it could be food, a threat, attacker, protector, playmate. If it is none of these at the moment it could become so (or has the potential to be). So the somatic value-link remains.

And it is from such discriminations related to us personally, to our basic needs, to the generalized second algorithm of our triune modular brain structure that all discriminations, syntheses, and the phonetics assigned to them derive their value to us. The value of any discrimination is in the relationship to us as coded to the second algorithmic representation of the genetic matrix. This is the basis of the word-value link or the fact-value link. The linkage is unbreakable. Although it can be externalized, and rarefied.

This linkage to the second, reciprocal algorithm of our evolved neurological structure provides the inevitable, unbreakable linkage between words, language, and ultimately fact and value that the positivists have tried so vainly and so distortingly to deny. The strength of the link is, of course, determined by the degree of importance or significance to the second algorithm or one of its myriad subroutines. The second algorithm gives the value anchor to any word, group of words, to any so-called fact. It answers the questions: Why is the fact important? Why do we want to know it? How do we feel about it? The value can range from great significance (like ***mama***) to simple awareness that anything unknown in our environment potentially threatens our survival (self-preservation).

Every word, fact, or sentence, then, is value-anchored. And it is significant as a word, fact, or sentence to the degree (direct proportion) of the strength of its value anchor. At the neurological level this value linkage is substantiated by findings that perceptual inputs from all five senses are processed through the limbic emotional centers where they are evaluated for long-term storage (as learning and memory) based on their significance to the various limbic structures.[94]

[93]This is very closely related to, but more fully delineated, than Damasio's somatic marker hypothesis (Damasio 1994). See Adolphs, et al., who building upon Damasio's concept hypothesize that context-dependent reasoning relies on feeling which is body-based (1996: 176). The extension here would be that context-dependent learning to include language would likewise be body-based on feeling.

[94]There is a substantial literature documenting the role of subcortical brain centers in learning and long term memory. For example see Aggleton (1992), which has numerous articles on

This will be covered in more detail in later chapters but it is increasingly clear that our discriminations are always, however seemingly indirectly, related to our survival/affection requirements or the second algorithm. This relationship may be at the level of the ego/empathy abstraction or at the level of one of the subroutines making up the self-preservation--affection master programs. All discriminations are linked, associated, related, or otherwise ***incorporated*** to the survival and affectional programming of our evolved neurological-somatic structure, and thereby, ultimately to the genetic informational matrix.

---

the role of the amygdala in emotion, memory and mental dysfunction. See also LeDoux (1996), MacLean (1990: 467-516), Donovan (1988: 198-201), Sarter and Markowitsch (1985).

Chapter 17

THE LIMITATIONS OF PRESENT DAY SYNTAX THEORY

It's tempting to say that imposing categories upon the world is one of the techniques employed by both animals and humans for the purpose of exploring it.
(Jean-Pierre Changeux, Neuroscientist) (Changeux and Connes 1995: 193)

Syntax theory, as currently articulated, cannot stand alone as a theory of grammar. The goal to achieve a syntax-only generative grammar that can produce all the well-formed and acceptable sentences in a language is doomed to failure. It is impossible on the basis of its own implicit assumption.

The basic assumption is that syntax is primary and that meaning emerges from proper syntax. Although syntax and meaning are highly interdependent, especially in written language, the causality, in fact, runs in the reverse direction. This mistaken assumption happened and was perpetuated by two factors:

a. The linguist observer sees the written sentence after the fact when lexical items bearing meaning have already been placed into syntax, or, in spoken language, as lexical items bearing meaning are in the ongoing process of being placed into syntax.[95]

[95] Recently, countering the mainstream of syntactic theory, there has been a renewal of interest in lexical semantics which investigates the relationship between the semantics of items in the lexicon and syntax. Although there is little consensus among researchers, it is usually held that lexical semantic representations of the predicates (verbs), in some cases, have a determining effect upon the syntactic expression of the associated arguments. See for

b. The preference for a mechanical theory in the sense of the natural sciences in order to be appropriately "scientific".

This chapter will examine both factors in more detail.

THE PRECEDENCE OF SYNTAX AND THE REVERSAL OF CAUSALITY: PRIOR ASSIGNED MEANING

In assigning syntax preeminence, linguists were misled by the overwhelming nature of the obvious. That is, by far most sentences they see or hear are ***serious*** efforts at communication (e.g., see Fiengo and May 1996; Braine 1990: 145; Grice 1975, 1978). The meaning is built in in advance as the meaningful parts are placed into syntax that perserves and, perhaps, extends that meaning. Such meaning assumes a common base of prior experience and understanding. An overwhelming percent of the sentences linguists or grammarians see are of this sort. And in such cases they become ungrammatical and sometimes ambiguous because of syntactical errors, especially in writing. The necessity for ***prior assigned meaning*** goes unnoticed until one sees an occasional outlandish sentence that contradicts our shared view of the world...from which we derive our shared meaning.

For example, the following two sentences are taken from Robert Borsley's introductory text on syntax theory

> Stefan succeeded in seeing Maja but he didn't see her.
>
> My lawnmower thinks I don't like it.

The author explained that the two sentences were unacceptable, although not ungrammatical. The first sentence was deemed unacceptable because it involved a contradiction. Although he did not say so, the author probably referred to the fact that the meaning of the first part of the sentence contradicted the meaning of the second part. The second sentence, according to the author, conflicted with our views of how the world is.[96]

---

example, see Levin and Hovav (1996), who, following a brief review of the literature and a case study, emphasize the value of seriously examining the hypothesis that the meaning of a verb determines its syntactical expression (1996: 505).

[96]Borsley (1991: 5). The divorce of syntax theory from meaning or semantics is glaringly apparent in Borsley's text. Neither meaning nor semantics is defined in his glossary nor listed independently in his index (only one item has the word *semantics* in it...a reference to Montague semantics under "M"). This rather astounding and artificial separation and denial of meaning in syntax theory reflects the pervasiveness of the implicit and ignored basic assumption. Meaning is covered by a lot of evasive, hidden, or disguised terminology throughout the text. And with good reason. If acknowledged it would destroy much of the argument and analysis.

In other words, both sentences, although not ungrammatical according to syntax, violate the implicit assumption of ***prior assigned meaning***. That is, we know from our commonly shared prior understanding of the meaning of the lexical items that, in the first case, one cannot both see and not see a person, and in the second, we share the knowledge (meaning) of lawnmowers as inanimate, nonthinking objects that cannot have likes or dislikes.

A similar example from Geoffrey Horrock's comprehensive text on generative grammar is the sentence

> The furious carrot slammed the door.

Horrocks explains that such a sentence

> ...may be rejected not so much on the basis of its linguistic properties as because the hearer cannot readily imagine a context in which the utterance in question could be used appropriately... (Horrocks 1987: 2-3).

Again, as in the case of the lawnmower, the sentence violates the implicit assumption of prior assigned meaning, which, of course, is always contextual. Such sentences, however, appear regularly in children's literature. They are playful and illogical statements that contradict a child's growing sense of meaning in context...to the child's delight. The children know they are being teased. By contrast, it strengthens their sense of reality and facilitates the growth of imagination and language. A child who has acquired the basic meaning of lawnmowers and carrots in a real life context is delighted by the playful challenge to its acquired sense of logic and meaning. Even sophisticated adults can still enjoy the charm and challenge of *Alice in Wonderland*. Only syntactically-focused linguists don't get it!

These above examples alone, and there are similar examples to be found in any text on syntax, reveal the implicit assumption that sentences will come with prior assigned meaning.

The sentences, although syntactically correct, are rejected as unacceptable in serious communication and not to be accepted for further syntactic analysis because they do not come with their expected prior assigned and taken-for-granted meanings. And you do not find examples of such sentences being given serious syntactical analysis. They are culled and tossed aside as soon as their defect is noted.

The foregoing analysis boils down to reversing the current precedence of syntax and leads to the precedence of meaning over syntax, and to the following conclusions:

Syntax can preserve or extend meaning, but it ***does not*** create it.

Faulty syntax can destroy meaning or make it ambiguous.

No amount of good syntax can create meaning out of nonsense or meaningless lexical items.

THE PREFERENCE FOR A MECHANICAL THEORY

The preference for a mechanical theory in the sense of the natural sciences also contributed to the precedence given syntax in theoretical linguistics.

After his usually taken-for-devastating critique of radical behaviorist B.F. Skinner's *Verbal Behavior* (1957)[97] on the primary basis of poverty of external stimuli, Chomsky postulated the necessity of an inborn grammar machine that produced language with only minimal external input. He called this machine a universal grammar and it followed the same mechanical operating assumptions of behaviorism, but moved a chunk of the mechanism inside the brain. The mechanism was still divorced from meaning and was to crank out syntax automatically, unthinkingly, or mindlessly.

This precedence of mechanical syntax has continued in Chomsky's theory and its offshoots ever since. Those offshoots that abandoned the concept of universal grammar, like GPSG, went even further in the direction of the precedence of mechanical syntax. Nevertheless, GPSG and other variants don't graph or analyze meaningless sentences either. Research continued as if there were an unspoken command hanging over it. That imperative could be expressed as

"Let's don't look for anything requiring anything as subjective as intelligence or significance. Let's look for and recognize only that which appears mechanical, mindless, and therefore objective."

It may be appropriate to observe at this point that objectivity or subjectivity can be emphasized separately for purposes of analysis. They cannot, however, be totally divorced, because of our inevitable subject/object relationship with the environment. They are used wrongly when they contaminate each other. Or when one is allowed to obscure the other.

EXTRA-LINGUISTICALLY-DRIVEN VARIANTS OF SYNTACTIC STRUCTURE

The purposeful focus in linguistics on purely "linguistic" factors is exemplified by the current state of syntactic theory. It has almost no place for semantics or meaning, but sees the rules of syntax as purely mechanically, independently generated. The failure to grasp or deal with the value linkage in language has been a serious weakness of linguistics as a science. It has led, in some cases, to an almost endless building up of rules for syntax and an almost equally endless building-up of tree diagrams to graph the rules, accompanied by a largely unsuccessful search for universals among the highly complex patterns.

[97]Chomsky's review of Skinner is reprinted in Geirsson and Losonsky 1996.

Although there *is* an innate algorithmic grammar, there is no specialized innate rote mechanical formula for the building up of the rules, conditions, and barriers, although you can nearly endlessly show variants on tree diagrams of sentences. Such variants are not produced by some infinitely complex inborn mechanical programming feature, mindlessly, as claimed by some writers.[98]

They are governed by two factors.

(1) They are are significance-driven by the value linkages to the second algorithmic matrix.

And in spoken language, but most conspicuously in the translation from spoken to written language

(2) Living-context-driven by

a. the qualifying power of vocal emphasis and body language

and by

b. the environmental context in which originally coded (learned) and also in which currently used.

SIGNIFICANCE

One of the most frequent criticisms of purely mechanical syntax is that there is no way to tell what is important. A common way to try to deal with this is by designating so-called lexical heads that dominate a sentence or portion thereof or assigning the behavioral effect of certain words to something vaguely called pragmatics.[99]

Take the two sentences:

Spot came quickly to John.

Death came quickly to John.

They are both syntactically correct and indistinguishable on the basis of syntax alone. But any normal human would recognize that there is an

[98]See Bickerton (1995: 39). Bickerton in his enthusiasm for the automatic mindless nature of syntax, devalues the importance of meaning. I think this is very distorting. Syntax is turned out rather habitually once the learned components are in place. But meaning is the entire essence of communication and determines whether a sentence makes sense or even communicates anything of value.

[99]There has been considerable recent work striving to clarify the definitions and boundaries, real or methodological, between semantics and pragmatics. For a review and discussion of some of the problems, see Matthews (1995). Matthews writes that pragmatics was commonly defined negatively as meaning minus semantics (1995: 51). Levinson (1995: 91) sees pragmatics as responsible for explaining how the same expressions may have different meanings in different contexts. Kempson (1996: 523) sees pragmatics, from the hearer perspective, as the study of interpretation of meaning from the psychological perspective. On the other hand, Lyons considers the separation between semantics and pragmatics as the product of methodological decisions and that the distinction does not necessarily reflect psychological or any other kind of reality (1995: 231).

enormous difference in significance. From the purely mechanical syntax we could even say

Snarth cone brickly to Jajf.

But we would have no idea what the significance could be.

THE LIVING CONTEXT VS. THE DIVORCED CONTEXT

In the living context of spoken language almost any element of the primary algorithmic framework of subject-verb-object can be moved anywhere (along with its vocally emphasized or body language designators, locators, and qualifiers) in spoken language. On the other hand, in the divorced from living context of written language, where we lack the designating, qualifying power of vocal emphasis and body language, the functions they perform must be substituted by formalized conventions. Such conventions are not innate but socially-produced and learned. And they can appear very complex. Without consideration of value-linked, semantically-driven, and living context factors, the process of sentence-making can truly appear to be a syntactically mechanical rather mindless cranking out of complexity.

The fact is that syntax is ***not*** created in a vacuum. Its basic framework is set innately by the process/framework of the primary algorithm, with other modifications tied to value, which is both innate and social, and to coded context and living context, which are always social.

Of course, the syntax-driven theorists can talk to artificial language programmers and cognitive psychologists because they, too, are often contextless and value-free, which is the reason their theories, as they pertain to natural languages, will always be either overly simplistic or mindlessly complex relying on infinite repetitions in logic and math of the primary algorithm without the second algorithmic value link. Without the link to value, and living context, the theories can never be psychologically adequate or emotionally satisfying.

Chapter 18

VALUE, CONTEXT, AND THE ORIGIN OF MEANING: SPOKEN VS. WRITTEN LANGUAGE

...Here we see a semantics which not only drains the concept of 'truth' of any moral content but divorces statements from their communicational context altogether.
(Roy Harris, General Linguistics, 1987: 159)

There are four parts to the ***meaning*** of a word or lexical item. They are, what I choose to call, the tag, category, significance, and context.

The ***tag*** is the name given any discrimination of any part of the environment. It is arbitrary to any language or group and may be a sound, symbol or both. It is objective to the extent that it is socially shared.

The ***category*** is the primary algorithmically based part of speech. The fundamental categories in English are noun and verb, reflecting the incorporating and extension, ordering and chaos aspects of the algorithm. All other categories such as modifiers and determiners are secondary. Categories, like tags, tend to be objective to the extent that they are socially shared in a language.

The other two parts, ***significance*** and ***context*** have objective aspects, but tend to be ***primarily highly subjective***.

Significance and context are the two aspects of meaning that will be further dealt with in this chapter. They are often overlooked, under-appreciated, or ignored in theoretical linguistics.

UNIVERSAL GRAMMAR AND THE PROBLEM OF MEANING

The universal grammar as proposed by Chomsky, along with its various offshoots, simply cannot account for the meaning of words. That is, how words get their meaning to begin with. It is not good enough in addition to syntax to just acknowledge and analyze an additional so-called semantic component. One must account for the origin and the strength of the meaning.

The origin and strength are found in how the sensory input (coded context) along with the phonetics and/or symbols associated with the words or utterances, are programmed in the framework of the primary algorithm to the subroutines of the second algorithmic matrix (e.g. keyed to somatic markers per Damasio 1994), and stored in long-term lexical memory. There are many considerations other than narrowly linguistic ones that go to explain this aspect of linguistics. The input of all five senses, visual, touch, sound, scent, taste may be programmed into the meaning of the words. All these accompanying sensory factors are encoded in part or whole to a word when it is learned and are available when the word is repeated.

This coded context is of fundamental significance with the basic vocabulary we get in our early childhood experiences. Words carry or suggest all these multiple sensory experiences plus their second algorithmic links of value and meaning (cf. Damasio 1994; Adolphs, et al. 1996). Our subsequent word accumulation and increasingly sophisticated, acquired syntax build upon and acquire meaning in part through extension and association with the basic language learned early in life.

THE NATURE OF SEMANTICS

The words ***semantics*** and ***meaning*** are generally taken to be synonymous or nearly so. There are two aspects to both, however, which are rarely if ever distinguished. There is ***objective meaning*** and ***subjective meaning***. The distinction is part and parcel of the well-known fact and value dichotomy. The linking of the two brain algorithms reveals that the separation of the two is artificial. Their linkage can be expressed in a spectrum as follows:

Objective ←----------------------------→ Subjective
Fact ←-------------------------------------→ Value
Primary algorithm ←-----------------→ Second algorithm

In our interpretations we may move closer to one extreme or the other, but the linkage is unbreakable. Moving too far in either direction only creates distortion. In a naturally, initially acquired language the two are blended. Any naturally acquired language is representational, however, and communication of meaning relies implicitly upon two key and largely unclarified assumptions. They are:

1) a commonly shared perceptual apparatus of the five senses coupled with a commonly shared cognitive/emotional substrate (which as assumed could only be innate or part of the protein expression of the genetic informational matrix; i.e. the primary and second algorithms).

2) a commonly shared set of basic life experiences in a commonly shared environment.

In a natural language environment these two work together to make social communication largely reliable. Even under the most closely matching circumstances, however, there will always be ambiguity because of genetic variation in the five senses, the wiring of the two brain algorithms, as well as differing individual experience within the same general environment. These are among the inescapable dilemmas and ambiguities of any social communication.

Present formal language semantic systems used in linguistics, e.g., the system devised by Richard Montague,[100] are, in their current state, inadequate to model or communicate meaning except in a purely objective, formal, or mathematical sense. They work according to the framework of the primary algorithm detached from the second algorithmic matrix. Such systems can deal effectively with the rather obvious, superficial, and somewhat tired old philosophical questions of so-called objective truth conditions, but without any link whatever to subjective values.

An example would be: John is a linguist.

Under a formal semantic system such a statement would be true only if there is such a thing as a linguist; and if so, that the person named John actually belongs to that group of things as claimed. This and other examples are rather obvious and trivial. For instance, the example of ***mama*** given earlier, with all its rich and value-loaded context, would simply be dealt with similarly as an objective truth condition.

Such simplistic, externalized, so-called objective "truth" conditions are completely inadequate in themselves to account for or to convey the texture and value of the word ***mama*** as originally learned by any normal child. They get at meaning in only the most superficial sense, missing all the power of living context, and survival and affectional significance that go with coding to the second algorithm of our triune brain structure (see the discussion by Harris 1978: 152-162).

The sentence, *That woman is my mama*, would be handled in the same simplistic manner. The formal semantical truth conditions or meaning would be satisfied if it is true that the individual specified is a woman, and, if being a woman, she is also a mother, and if being a mother, she is also, specifically, the mother of me. Oxford professor of general linguistics, Roy Harris, sees this over objectification as the outcome of what he calls the "myth of the language machine" (Harris 1987: 162). In referring to

[100]Montague (1970). See the introductory work on Montague by Dowty, Wall, and Peters (1981).

semantics along the line of A. Tarski, whose general semantic approach Montague also follows, he writes:

> ...Here we see a semantics which not only drains the concept of "truth" of any moral content but divorces statements from their communicational context altogether (Harris 1987: 159).

In the apparent effort to penetrate deeper into meaning, the procedure is devised of assigning numerical or symbolic weights to represent the value of variables. This, however, ***relies on a prior subjective interpretation of the value***. Such techniques can only at best communicate the message..."I assign more or less significance to this word, concept, or variable, from my interpretation of its meaning, whatever the heck it may mean or signify to others."[101]

Of course, since all languages are representational, even naturally, initially acquired ones, logical and even machine languages of strings of 1's and 0's could conceivably come to communicate subjective or second algorithmic value. They could achieve this, like any later acquired second natural language, by the reinforced association, word by word pairing with the naturally-acquired language. The subjective value component of meaning would then be by derivation from the original language. The words, "I love you," ***could*** carry a derived power punch of significance, even if read in long strings of machine code, but speaking such strings may be impractical, tedious, and test the limits of short term memory in any significant communication.

[101]The superficialities and inadequacies of traditional formal semantic theory have been recognized by a number of recent scholars. Linguist Ray Jackendorf (1996), building upon a distinction by Chomsky between external semantics and internal semantics, discusses the framework of an enriched semantic approach, which he calls ***I*** (for internal) ***semantics***. I-semantics, unlike the external approach of formal semantics, attempts to connect with the more general issues of psychology, neuroscience, and biology. Although admittedly in its infancy, I-semantics sets itself the goal of connecting with cognition. Concerning the problem of meaning, Jackendoff poses two basic questions: The first concerns the terms in which humans grasp the world, i.e., what are the formal and substantive properties of human concepts/thoughts/ideas? The second asks how these properties may be formalized in order to develop a fully explicit and predictive theory? Although this approach focuses on the connection with cognition, if it is expanded to include the emotional components of survival and affective neuroscience, it may position itself to begin dealing with the fundamental problem of value and significance. There is also a group of researchers who strive to move past the static interpretation tied to the classical mathematical logic and set theory of formal semantics toward a dynamic interpretation that equates the meaning of a sentence to its potential to change information states (for a discussion see Groenendijk, et al. 1996; cf. Bartsch 1998). These so-called dynamic interpretations still, however, do not engage the critical question of subjective significance or value to which any meaningful human communication must be anchored.

THE TEXTURE OF THOUGHT: FURTHER PROBLEMS OF OVERSIMPLIFIED OBJECTIFICATION

Some linguist authors, who see language as a faculty separate from other aspects of brain function and cognition, ridicule the idea that people also think in images.[102] Such a position should be considered overly limiting and "linguisto-centric."

All our senses are associated to some degree with words, especially the early ones that we learn. And it is largely upon these early words with intense significance and value which our later sophisticated semantics or semantical connections are built into ever higher and more abstract vocabulary and sentence formation. The basic words of language are learned and carried with all the sensory information--to include the visual pictures of them that we saw and associated with them.

Yes, we ***do*** think in images. We do not think solely in completely detached linguistic entities called words. We think in images that we got coded in our neurological structure through our visual perceptual apparatus. We also think in sounds, smells, tastes, and touch. Sometimes it is very difficult to put such complex thought associations of words into even spoken language with all the variation of vocal emphasis and body language available to us.

How much more difficult it is to compress the meaning, associations, and values into the forced restrictive conventions of written language, along with the visual and other sensory data associated with the pictures we have in our heads. We have the coded context of words and pictures and other sensory data, plus a value link with our second algorithmic matrix, all associated together. Once they are associated we invariably get them all, in part or whole, subliminally or consciously, when we again hear or recall the words.

The power of poetry and fiction depend on these associations, draw upon them, and would be of no effect in their absence.

Totally externalized, objectified systems, such as Generalized Phrase Structure Grammar (GPSG), which recognize no necessary connection with psychological reality, can get extremely complicated and mathematical but they are unable to hit the core of natural human language. They fail because they have no link to an evolved informational matrix of survival and bonding, of meaning and value. GPSG may truly be useful in directly working with computer programmers and cognitive psychologists, because the latter also operate without such a value connection (see Gazdar, et al. 1985).

Lacking such an evolved value core and failing to grasp coded context, computer scientists, programming for artificial intelligence, must duplicate

[102]Bickerton (1995: 22-24). For a contrasting view see Pinker (1994: esp. pp. 70-73).

or mimic and program for all the contingencies of syntax. They must work with the mathematically complicated and sometimes seemingly endlessly complex sentence and phrase structures of human language with no effective way to tell what is important and what is not.

Some of the value assumptions, however, do work into the programming because, although the linguistic models and the computer models that are employed lack value links, the human beings doing the work don't lack such links.

HOW WORDS HOOK ON TO THE WORLD

For the better part of a half-century, philosophers of the linguistic-analytic persuasion pursued the question of how words hook on to the world. The quest, although producing much published work and some valuable insights, generally failed to answer the question.

In the positivist tradition, language was viewed almost entirely in its immediate social context. This overly inclusive, immediate social focus was inadequate as it also proved to be in radical behaviorism. Such a focus overlooked or ignored the fact that humans were born into their immediate society bearing innate mechanisms for thought and language which had evolved by selective effect in the social environments or contexts of the evolutionary past. They brought these mechanisms of the evolutionary social past into the immediate social present.

Contributing to the failure to answer the question how words hook on to the world, then, was the lack of an adequate conception of how words hook on to the individual human. The corrective to this overly social perspective was initiated largely in linguistics by Noam Chomsky with his concept of grammars innate to the human brain.

Since linguistic philosophers and academic linguists often go their separate ways, Chomsky's approach which must rest ultimately on points of connection with neuroscience, had surprisingly little effect on the philosophical linguistic quest. Rorty, in his *Philosophy and the Mirror of Nature* (1979), recognizes this and gives concession to the possibility of prewired, neurological subroutines bearing on the issues of language (Rorty 1979: 241).

Understanding that words hook on to the brain by way of the framework and process (syntax) of the primary algorithm "hooked" to the master and subroutines of the second, reciprocal somatic brain algorithm, carrying with them elements of the entire social and physical environmental context mediated by our five perceptual apparati (which are also in part coded with the learned word), can take us further along the road to understanding the rather complex basic question.

Chapter 19

UNIVERSAL GRAMMAR AND THE EVOLUTION OF LANGUAGE

...the full set of such principles constitutes a formal grammar, which can be seen indifferently, as having an analytic or parsing function, or synthetic or generative function.
(Karttunen and Zwicky, Linguists, 1985:3)

From the standpoint of origins, all human languages can be seen as having three components: Firstly, they have a universal, innate component prescribed in the genetic informational matrix, which, translated into protein behavioral structure, provides the basic structure and framework. Secondly, there is a socially-evolved component which draws on and builds upon the universal. Thirdly, there is a formal or stylistic component which is a part of and extends the socially-evolved component (cf. Jackendoff 1994: 34-35).

THE UNIVERSAL COMPONENT OF LANGUAGE AND THOUGHT

Simply put, we ***do*** have a universal grammar. It is not, however, distinctly and purely linguistic as claimed by some scholars. Language, like thought, expresses the primary algorithm of organic synthesis and the reciprocal algorithm of our triune modular brain structure, which represents the actualized somatic matrix. That is, it grows out of our subject/object relationship with the environment as driven by the requirements of our self-preservation and affection programming. The two algorithms provide the innate syntax and the innate semantic value to our universal grammar.[103]

[103]Pinker (1994: 81-82) in writing about "mentalese", his name for the universal grammar,

The order of our innate syntax is that of the primary algorithm...the innate structuring, incorporating, synthesizing algorithm that is easily identifiable in many variant expressions across the range of human thought and behavior.[104] The elements of ***actor-action-recipient***, ***subject-verb-object***, or according to Braine, ***argument-predicate-argument*** (1994: 25-27), reflect the algorithmic components of ***order-chaos-order***, ***incorporation-extension-incorporation***, ***object-relationship-object,*** etc. This component assures self-reference...that we are the subject of all our sentences. Verbs express the relationship between us and all objects or discriminations in our environment.

---

sees it as a language of thought, on one hand richer than any given language and on the other hand simpler. For instance mentalese would not have conversation-specific words like the determiners "a" and "the" or information about pronouncing them or ordering them.

[104]Even though it is not recognized or identified as such, the primary algorithm bleeds through in much of the contemporary writing on the relationship between semantics and syntax. As one example, Kempson (1996) makes her case for a natural language as a labelled ***deductive*** process from the perspective of the ***hearer.*** Deduction, of course, proceeds algorithmically from the general to the particular, from the the whole to its parts. But the hearer perspective is only one side of the language process. From the other side, the ***speaker*** side, which Kempson ignores, her position implies necessarily that the speaker ***induces*** or assembles the intended meaning of the communicated expression from lexical or grammatical parts. In the two-way communication process, then, the speaker's role is ***inductive***; the hearer's is ***deductive***. Therefore, in the necessary roles of speaker and hearer, common to all human language communication, we can see the inductive/deductive mirror image of the primary algorithm. When Kempson adds movement to the constructive/deconstructive (interpreting) process, she is, of course, expressing the primary algorithm as a dialectic in the manner of Piaget's dialectic of constructive equilibration, which is also reversible into its parts. Although the elements of the primary algorithm are clearly present, Kempson's presentation is unclear. She speaks of building up or constructing the meaning of an expression ***deductively***. It is more accurate, however, to think of building up or constructing as ***inductive***...the act of synthesizing, the assembling of parts into a whole, the movement from particulars to the general. On the other hand, ***deduction*** is better thought of as the analyzing function, the breaking down of a whole into its parts, moving from the general to the particular. At the conclusion of her argument, however, she writes that her model "suggests that the ***first step*** (emphasis mine) in the assignment of an interpretation to natural language strings is the incremental inferential procedure, whereby as a part of the ***parsing*** (emphasis mine) process a hearer ***creates*** (emphasis mine) a configuration which represents some propositional content."(1996: 596). Perhaps Kempson is implying more than she says. Here, in her own words, she conflates the processes of parsing (analysis, deduction) and creating (generating, synthesis, induction) into one first step (cf., Dowty, et al 1985: 3 and p. 158, this text). Perhaps she ***implies,*** or even ***intends***, a two-step (rather than one-step) deductive/ inductive interpretation process in which the hearer breaks down or ***parses*** the elements of the expression ***deductively*** and then reassembles or ***generates*** the interpretation ***inductively***. Despite the ambiguity, the framework of the primary algorithm, nevertheless, comes through clearly.

The value (emotional meaning) of individual words and groups of words is derived from coding to the subroutines of the second algorithmic matrix. The strength of the meaning or value link is in proportion to its significance to our basic subroutines of needs. That is, words or concepts having great self-preservation or affectional significance to us carry a greater charge of value than do those carrying less significance.

THE SOCIALLY-EVOLVED CORE COMPONENT

As spoken language first begins among human groups, sounds are uttered from among the range possible to the human vocal structure. The relationship among the sounds, now become words representing objects, happenings, and other discriminations, was maintained in simple algorithmic integrity by changes in vocal emphasis and by body language, to include pointing, gestures, facial expressions, etc., within the living context of social communication.[105] Each language, then, could be seen to have produced a socially-evolved core component.

As languages became more sophisticated, vocal emphasis and body language were replaced by sounds (words) which designated objects and direction of action. These sounds (say, articles and prepositions) attached to the objects and verbs to maintain ***algorithmic integrity*** or the relationship among parts previously provided by vocal emphasis and gestures. This development of these designating, indicating particles was of great importance to the emergence and development of written language because they had to preserve meaning divorced from the living context of vocal emphasis and body language.

The basic grammatical parts of language (what we name subject, object, verb) could then be shifted around in any way as long as these designators and direction of action markers went with them or in some other manner maintained their reference.

THE FORMAL, STYLISTIC COMPONENT

As written language further evolved, each different human language developed its own, more or less unique, add-on rules of syntax to maintain the relationship among the critical parts...to maintain algorithmic integrity.

These socially produced rules, which extended the socially-evolved core component could be quite varied and complex. Some of them were unnecessary to meaningful communication, but became matters of preferred style. Examples in English would be the prohibition against split infinitives and the admonition not to end sentences with a preposition.

[105]Gomez (1998) sees this ostensible communication, seen likewise in the great apes, as more fundamental and powerful.

The contrast between this stylistic component and the more basic algorithmic component of language can still be seen in the use of telegraphic style, where only the minimal language is used to communicate.

An example of telegraphic style was burned into my youthful neural circuitry when I was a lad of about ten years. The year was 1942, in the beginning of World War II. We had just moved to another city to support the war effort. My mother had put our family home up for sale with a lawyer/broker in our hometown. Since housing was much in demand, she shortly received a telegram from the lawyer. It read:

> I have a sale for the property of Mrs. Alice C. Cope. The prospective buyer, who has recently transferred into the area, proposes an offer amounting to the total sum of two-thousand dollars. He will provide the proposed sum in the form of cash as payment in full for the property. Please let me know by return telegram whether you wish to accept or reject this offer. Cordially, I await your response.

This telegram came collect and I remember that my mother was furious. She complained to my grandmother who was living with us, that the wordy and expensive (in those days) telegram could have been easily composed in the ten word minimum charge range. To make her point, she composed the following version and read it to my grandmother, who of course nodded in agreement. I simply listened awestruck.

> Have offer house. Two-thousand dollars cash.
> Wire accept, reject.

Although I did not realize it at the time, her version was pure primary algorithmic "bare bones". It had only nouns and verbs. No stylistic components. And it communicated perfectly, because the social context was mutually understood. The minimal primary algorithmic process is also seen in the development of pidgin and creole languages. Such languages are developed by people from different linguistic backgrounds who are thrown together with no common means to communicate. They soon develop a language that is similarly "bare-bones" primary algorithmic. The stylistic components are added in as the language develops (see the discussion in Pinker 1994: 32-39).

GENERATIVE GRAMMARS

In linguistic theory, generative grammars are defined as grammars capable of generating all grammatical sentences of a specific language.[106] This is a formidable task indeed. Such grammars must deal successfully with all three components, two of which are unique to the specific language

[106]Such grammars are called observationally adequate. To be fully adequate such grammars must be descriptively adequate, assign a structure to each well-formed sentence. To be explanatorily adequate, a theory must select the best available descriptively adequate grammar (Horrocks 1987: 14-23).

and may have no applicability across the spectrum of human language families.

The rules of such generative grammars will tend to get more and more complex as we move from the universal component, through the socially-evolved, and the formal stylistic levels. In the latter two levels the patterns identified will not be universal patterns. They may at best be general patterns if they are found to apply to languages other than the specific one being investigated.

Nevertheless, underpinning them all lies the innate universal syntactic framework of the primary algorithm and the coded value-link to the second algorithmic matrix. As noted earlier, the hierarchical tree is the pervasive method for graphing the rules of grammar. And as David Harel of the Weizmann Institute observed in his text on algorithms in computer science, trees with their numerous and varied applications...

> more than any other structural method for combining parts into a whole...can be said to represent the special kind of structuring that abounds in algorithmics (Harel 1987: 43-44).

The primary algorithm is expressed structurally and functionally in generative grammar. The incorporation, association, synthesis, or ordering of parts into language wholes is rendered structurally by the hierarchical tree. The process, however, is functional. The nature of the primary algorithm is double-faceted...it is both structure and function, framework and process. In this synthesizing or ordering of linguistic entities into sentences, generative grammars clearly follow the framework and process of the primary algorithm.

In prior chapters the primary algorithm, with its link to the second, has been demonstrated to underpin and guide the flow and structure of human thought, subjective experience, and behavior. Small wonder that it also guides and expresses the flow and structure of language. For language is none other than the uniquely human, monumentally powerful, spoken and written vehicle for the social communication of our thinking, our experiencing, and our behaving.

INHERENT AMBIGUITY AND THE PITFALL OF THE OVER-SERIOUS GENERALIZATION

In closing this chapter, it is important to point out a recurring pitfall that results from the inevitable linkage of language with the algorithmic patterns of our thought. The inherent ambiguity of the discriminating, generalizing process was noted in the chapter on contemporary philosophy. It is a seductive pitfall--in science and philosophy--to overlook this ambiguity. And this same seductive tendency manifests itself in current linguistic theory.

In the preface of his book, *The Language Machine*, Roy Harris of Oxford University, sees what he calls the myth of the language machine having its roots in the post-Renaissance European culture which drew its original psychological support from the momentous invention of printing. According to Harris, the myth sees languages as fixed codes for thought transference among human beings and provides a social institution of language to make such transference possible. Harris complains that it is a myth which ignores differences between individuals in favor of emphasizing collective conformities. In doing so, Harris asks how linguistic unity arises from linguistic diversity:

> Since from cradle to grave, the personal linguistic history of every individual is unique, how is it possible that this rich variety of linguistic experience at the individual level should ever give rise to a common language of the kind which the myth postulates? (Harris 1987: 7).

Harris sees the mythical language problem demanding the mythical solution of a biological language machine that automatically constructs communal linguistic systems fully-equipped to solve all the basic communication problems of human survival (Harris 1987: 7-8).

The issue Harris raises is the cautionary one inherent in the nature of the primary algorithmic process of generalization. Generalizations always obscure differences. They obscure differences, contradictions, in the relationships among the parts which never fit exactly, but only generally. That's why they are called generalizations. And that's why we must not take them too seriously. When we take them overly seriously, and when the differences or contradictions they obscure are truly important ones, like the communication of meaning, values, and significance, such generalizations can become damaging to the cause of science as well as to the substance of humanity.

Generalizations taken too seriously become belief systems or myths. We are used to seeing them in religious or historical ethnocentric belief systems. There is also, however, the scientific version...that is, the scientific operational research position of convenience become taken as reality. The differences and contradictions such positions ignore for research purposes, become the differences and contradictions they deny in application to reality.

The most egregious and notorious of the scientific versions is the myth of radical behaviorism, which was so influential during the mid two quarters of the 20th century. Scientific behaviorism was underpinned by the experimentally useful but narrowly defined and focused stimulus-response model, and B. F. Skinner's later operant conditioning. The radical behaviorist research position, by obscuring, if not outright denying, the enormous, active, dynamic genetic programming of the human organism, allowed Skinner to extrapolate the narrow premises into an ***explanation of everything***...and a prescription for an ideal society that denied the necessity for freedom and dignity...those natural, genetically-driven, necessary and inevitable, as well as all-important, expressions of our higher brain development (Skinner 1970).

The danger of such scientific generalizing from limited, although perhaps legitimate, operational research positions of convenience, is that some dictator or group of well-meaning technocrats might actually try to impose such a system. The behavioral tension such imposition would create by its denial of freedom, dignity, and the genetically dynamic nature of the human organism would lead ultimately to more mind control, oppression, and ultimately revolution.

The inherent logic of behaviorist's ***environment is everything*** position is that any failure of the program results from the failure to properly control environmental contingencies. This deadly logic, based on the faulty environmental premise, leads inevitably to increasingly oppressive control in response to resistance to or failures of the program.

The former Soviet Union could, perhaps, provide an imperfect example of such a program. Freewheeling democracies can quickly render such programs ineffective and throw them out. A technocracy, however, convinced of its science, and with the power and resources to implement it, might not. Inherently ambiguous generalizations, then, when taken too seriously, become rigid. They then become mythic and arbitrary, defensive, and ultimately oppressive in the process.

Current linguistic theory has never come adequately to terms with the issues of meaning and value. Where the theorists have not denied the contradictions, they have often ignored them as unimportant. Worse yet, when they have tried to deal with them, they have forced the differences and contradictions to fit the currently inadequate theoretical generalization. They have distorted and trivialized meaning and value by over-objectification.

As currently conceived, the language machine, is a caricature expressing the primary cognitive algorithm divorced from the second algorithm of our triune, modular brain structure and, ultimately from the genetic somatic informational matrix. The contradictions, the differences, the caricature obscures, and the tension that accompanies such obscuring, can only be resolved when the somatic algorithmic link is joined in the paradigm.

Chapter 20

THE PRIMARY ALGORITHM AND CHOMSKY'S MINIMALIST PROGRAM

Concepts of fractal geometry are most useful
in characterizing the structure of branching trees.
(E. R Weibel, Anatomist, 1991: L361)

In recent years Chomsky has reexamined his explorations into universal grammar and the phrase structure grammar that derived from its assumptions and premises. His new minimalist program abandons the former rule based system for one designated as principles and parameters (P&P). Language is seen, then, not as based upon rules of syntax or phrase structure, but upon universal principles that are constrained by a finite number of options designated as parameters.

This is almost like going back to square one, but not quite. In its abandonment of the search for a rule-based universal grammar, P&P retains the enormous amount of data generated by the research of recent decades spanning many languages. There are more data to work with now than before. Chomsky sees the tasks at hand under the new minimalist program more difficult and more interesting. In his own words

> The primary one (task) is to show that the apparent richness and diversity of linguistic phenomena is illusory and epiphenomenal, the result of interaction of fixed principles under slightly varying conditions (Chomsky 1995: esp. the introduction and chapter 4).

THE MINIMALIST STRUCTURE

The structure of Chomsky's proposed language faculty runs roughly as follows.

The faculty consists of a cognitive system (CS) and two performance systems. The CS interacts with the performance systems which are designated the Articulatory-perceptual system (A-P) and the Conceptual - intentional system (C-I).

CS interaction with the A-P system produces what is called Phonetic Form (PF), which would be the sounds attached to the utterances.

CS interaction with the C-I system produces what is called Logical Form (LF), which would be the structure of the utterances that conveys the message.

The CS is itself composed of two parts, a computational system, and a lexical system. The computational system draws it words from the lexical system and applies its operations to them in order .to produce human language. The computation of human language is abbreviated as CHL.

The structure can be expressed in abbreviated form indicating a possible bi-directional flow between the cognitive system and the performance systems:

CHL=CS (COMPSYS+LEXSYS)<---->input/output <----> C-I + A-P

The minimalist P&P conceptual approach permits a better demonstration of how the primary algorithm shapes language than the generative grammar approaches of recent years. We can, in fact, say that within the cognitive system (CS), the computational system applies the primary algorithmic structuring process to the lexical items drawn from the lexical system to produce the language output. This statement can be clarified further by looking closely at the lexical system.

THE PRIMARY ALGORITHMIC STRUCTURE OF THE LEXICAL SYSTEM

The lexical items, themselves, have been stored in memory according to the algorithmic process. The fundamental categories of lexical items are defined by their algorithmic function. There are only two. They are what are traditionally called in English grammatical terms, ***nouns*** and ***verbs.*** The category of an incorporated item is one of four parts that make up its meaning: the other three parts being its tag, its context and its significance.

Nouns. All objects and concepts that can be discriminated or constructed by our perceptual/cognitive system are given ordering, incorporating (into themselves and into us) labels, or rather, tags. These tags (words, names), which may be phonetic and/or visual, are arbitrary and they are the hooks by which the discriminated objects are filed (made a part of, incorporated into the lexical system).

Objects are not only given tags arbitrarily, they are also selected arbitrarily out of the total texture of the environment...by the algorithms of our perception. Objects are all synthetic, in that, they are parts taken as a whole, either willfully arbitrarily or because our perceptual system cannot discriminate them further into parts.

Verbs. All extensions beyond the bounds of an object or concept, relating the internal parts of an object or concept, relating the object or concept to other objects and concepts, or relating objects and concepts to the general undifferentiated environment are covered in English by labels and are called verbs (conjunctions and prepositions also perform limited extension or relating functions). In other languages they are called differently. In Japanese, for example, verbs are called *doshi*, or words of movement. The verb "to be", for instance, relates the object to the general undifferentiated environment. All other verbs, representing the extension, chaotic aspect of the algorithm, express movement, relationship (the resolving of chaos), change.

Nouns and verbs (algorithmic discrimination/orderings and extensions) are the fundamental algorithmic categories that are the understructure of language. All other lexical items qualify, quantify, or designate such orderings or extensions. These other lexical items are the traditional adjectives, adverbs, and particles. They are not fundamental, but secondary. And in language they hang with the fundamental categories they qualify, quantify, or designate.

The lexical items, then, are so to speak, pre-cut. That is, they have their preassigned substantive category (noun-object, verb-relational), for the algorithmic process. Although in some languages like English, they may easily be converted to the alternative category by conventional rules. Almost any verb can be converted to a noun by adding an appropriate suffix like, ...ing, ...ion, et al. And many nouns can function as verbs by adding such suffixes as ...ize, ...u(i)late.

Given the algorithmic category established by the lexicon, the computational system, then, must function to ***preserve*** algorithmic integrity when adding in the secondary categories in order to grind out the sentences of language. The computational system must, accordingly, operate necessarily on the same primary algorithmic basis as the lexicon in order to do its job. Its job is to apply, check, and preserve algorithmic structure in its outputs to the articulatory system (whether spoken, signing, or written).

The computational system must assure that such outputs match the similar algorithmic protocols in the brains of the humans for whom the messages are intended. Algorithmic integrity must also be maintained in keeping with the socially evolved and agreed sounds, symbols, and conventions of the particular language used. Otherwise, confusion, ambiguity and a breakdown in communication occurs. That is...people will not be able to tell who did what to whom.

THE INFINITE USE OF FINITE MEANS: THE MANDELBROT ANALOGY

Language has often been described as a system characterized by "the infinite use of finite means."[107]

The structure and function of the primary algorithm allows for this phenomenon. In its simplest form the primary algorithm can be represented structurally as the simple and basic form of the hierarchical binary tree.

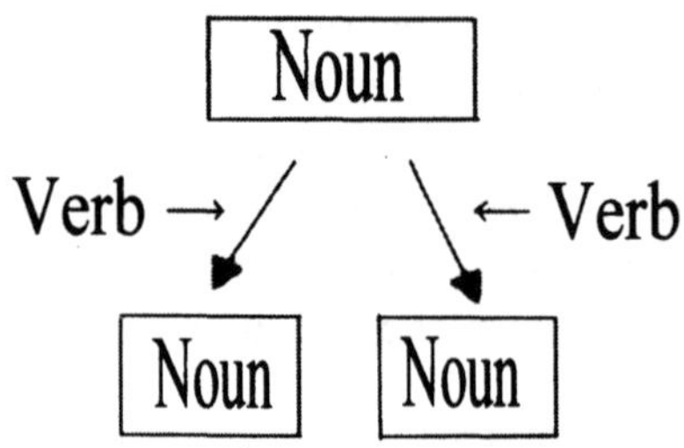

Figure 20-1. Hierarchical tree showing basic parts of speech.

The nodes represent the objects (nouns without names). The bi-directional lines represent the extension, movement, relationship (unnamed verb function). This simple algorithm (perhaps, but not necessarily, expressed with more than two nodes), repeated and interconnected hierarchically and recursively can produce limitless variety and complexity. In the case of language, it can represent ever finer as well as ever grosser discriminations and orderings, depending on your direction of travel in the interconnected hierarchies.

Here is also the algorithmic origin of the "level of generalization" or "level of analysis" issue in any discipline. How you view things depends on at what level you cut into the hierarchical organization.

An interesting and perhaps appropriate analogy to the cognitive primary algorithm may be found in the fractal geometry of Mandelbrot in mathematics. Fractals are geometric designs that appear as infinitely smaller or larger (depending on your perspective) repetitions of the same shape. They deal with the geometry of hierarchies and random processes. Fractals are pervasive throughout nature, specifying the shapes of snowflakes, coastlines of continents, etc. Such algorithmic patterns are so pervasive in

[107]The saying goes back to Humboldt (1896/1972), who said that language makes infinite use of finite media.

nature that identification of them has led to the technology of data compression in the computer industry.

The primary algorithm is analogous to, if not an actual biological variety of a fractal algorithm. A number of researchers have applied the principles and concepts of fractal geometry to biological systems. Swiss anatomist, Ewald R. Weibel reports that concepts of fractal geometry are most useful in characterizing the structure of branching trees. Anatomically, he finds such trees in airways of the lungs and in blood vessels (Weibel 1991: L361-369).[108]

The biological growth pattern from neuron to successively higher hierarchical nets of neurons may be viewed similarly. The growth pattern indicates that the basic structure reaches a built-in or environmentally-imposed constraint, branches, and regenerates its basic form repetitively at increasingly higher hierarchical levels (e.g., Ball 1988 applies fractal scaling to neuronal dendritic growth; see also Lowenstein 1999 on the closure of molecular loops). Bruce West, who discusses the development of a fractal theory of medicine, argues that the organization and diversity, seen at all levels of anatomy and control processes in human beings, should be described in terms of fractal processes (1996: 291).

Biological fractals, however, like the primary algorithm, are more complex than the standard fractals of the physical non-living world. In addition to the seemingly infinite pattern repetition, the primary algorithm is characterized by cybernetic feedback, feedforward loops, and general recursiveness characteristic of the protein-nucleic synthesizing loop, which, in turn, rests ultimately on molecular configuration and the four binding forces of physics.

One of the most conspicuous features of texts on syntax is the seeming endless strings of hierarchical trees of phrase diagrams, recursively, repetitively, imbedded within each other. Even a superficial inspection of these strings of trees suggests their fractal nature and such seems necessary to their coherence. Until the fractal-like nature of the primary algorithm is recognized as such, linguistic researchers will continue to be lost and wander endlessly in analysis among the potentially infinite, detailed and elaborate intricacies of its repetitive expressions.[109]

The entire basic vocabulary (lexicon) of any human language consists of finer and finer or grosser and grosser repetitions of the basic elements of the

[108]Havlin, et al.(1995) and Korolev, et al. (1994) apply fractal analysis to DNA sequences. Dewey (1994) applies them to biopolymers. Also see Dewey and Spencer (1991) for an article looking at the fractal dynamics of proteins.

[109]See Van Loocke (1999) for a proposal of a type of growing fractal with relevance for biology. For additional discussions on fractals as applicable to biological systems, see Mac Cormac and Stamenov (1996); Barnsley (1993).

primary algorithm. It is for this reason, and not to lose the attention of the reader, to the not so obvious repetitive vocabulary of the primary algorithm that I have repeated the words ***order, incorporate, synthesize***, ad infinitum for the incorporative aspect of the algorithm as well as ***extension, chaos, movement, change***, ad infinitum for the extension aspect of the algorithm.

Other than the variety derived from giving all objects that can be discriminated a separate tag (ordering or incorporation), our entire vocabulary is made up of algorithmic repetitions with minor discrimination differentials. All objects, concrete and abstract, that we name are products of our discrimination (analysis and taking apart) and ordering (labeling and making into wholes, units, objects). It is a survival mechanism. When we discriminate, then label or order, we identify and place the object within our environment and in that way achieve some control over it.

A close inspection of any dictionary will reveal the repetitiveness of the basic lexicon. We make endless minor discriminations--perhaps to avoid acknowledging the fact that we are saying essentially the same thing again and again. This gives the illusion of great richness and variety, which also serves, perhaps, the purpose of entertaining and stimulating that magnificent organ called the brain and not boring it to death.

Although the entire lexicon is produced by and according to the primary algorithmic process, it is enlightening to catalogue how many actual words relate to the algorithm itself. For instance, all synonyms, antonyms, modifications, extensions, derivations, and deviations of the words *order, incorporation*, and *synthesis* cover a sizable portion of the vocabulary we use. Some examples can be found on almost any page of the dictionary. A short list is furnished in appendix two at the end of this book. Readers can surely add to it by a dictionary search of their own.

FROM LINGUISTICS TO NEUROSCIENCE

The prewired, genetic aspect of language is an empirical question that must ultimately rest not on simplistic reduction, but upon links or points of connection, to neuroscience. Neuroscience is the appropriate level to study the genetic and environmentally-constrained substrate which makes the development of language possible. It is important to keep in mind that the development of language is not inevitable. Language depends for its development on a social environment. Its development appears inevitable only because humans survive almost inevitably in a social environment.[110]

I turn in the next chapters to the specific neurological mechanisms that, to the best of our current neuroscience, are known to contribute to our almost

[110]For evidence of the necessity for social interaction is language acquisition see Snow (1977). Also there is evidence for critical periods of social exposure to the language acquired. See Lenneberg (1967), Newport and Suppala (1987), Johnson and Newport (1989).

inevitably realized potential for language development. We should not expect, then, to find an innate, isolatible, fully-assembled neurological machine dedicated to the sole purpose of mechanically cranking out language. We should expect to find neural circuitry, with some areas of specialized function interconnected to all the areas that affect perception, self- preservation and the affectional capacity to nurture and sustain social life.

It is the province of neuroscience to study how the neurons and networks of neurons are formed and how they operate to provide us with the potential to understand, use, and experience our thought, emotions, and language in this way.

Chapter 21

LANGUAGE, BRAIN, AND NEURON

Brain-Language coevolution has significantly restructured cognition from top-down.
(Terrence W. Deacon, Biological anthropologist, 1997: 416-417)

The capacity for language, spoken and written, rests upon a neurological substrate that has both structure and function and which is ultimately somatically bound to the genetic informational matrix through behaviorally active proteins. To be meaningful language must also be linked to thought. If not linked to thought we would have nothing much to communicate, except the basic repertoire of lower mammals...warning cries, separation cries, comfort and attachment vocalizations, et al.

In previous chapters, I explored the algorithmic links between subjective experience, behavior, thought, and language. The primary algorithm was shown to be the unifying, pervasive cybernetic framework and process algorithm across the spectrum of these analytically distinct aspects of human function. The engagement of the second algorithmic somatic matrix, the master programs of self-preservation and affection, ego and empathy, was shown to be necessary for subjective experience and the emotional charge or significance of thought and of the learned lexical items of language.

Given the linkage of primary algorithmic higher cognitive centers with the second algorithmic emotional and motivation centers of our triune modular brain structure, we should expect to find the neurological substrate for language widely distributed within the brain. Evidence from neuroscience indicates increasingly that this is so.

WIDELY DISTRIBUTED OR ONE MODULE-ONE FUNCTION

There are currently two main, roughly competitive, viewpoints in the study of brain organization for language. First is the viewpoint of widely distributed function referred to above. The second is a modular viewpoint that in its extreme form postulates one module for each function.

The one-module-one-function concept is a convenient and somewhat natural way to think of structure. This is especially true, for instance, when one is designing machines, like computers. It is a hypothesis of convenience which apparently has some reality in actual neural structure. For example our processing of visual stimuli is known to take place mainly in a specific area of the brain. The visual cortex is not however an isolated module. It is extensively interconnected to other areas of the brain, down through the subcortical areas to include the basal ganglia and the brain stem.[111]

Certain areas of the brain seem to have circuitry dedicated specifically to some language functions. Broca's and Wernicke's areas have long been identified as areas of language function. Broca's area, located in the left cerebral hemisphere, in the rear of the frontal cortex, was identified by the French surgeon and anthropologist, Paul Broca (1824-80). In his studies of aphasia (speech loss), Broca found that his patients had suffered some damage to the area of the brain which later came to bear his name. Wernicke's area, located to the rear of the left cerebral hemisphere, near the auditory associational area, was identified by the German scientist, Karl Wernicke (1848-1905), in the later part of the last century. Damage to this area results in an inability to fully comprehend spoken language. Speech production itself is not necessarily impaired, but choice of words is often confused and inappropriate.

More recently, research has shown that damage to the circuits connecting Broca's and Wernicke's areas results in a condition called conduction aphasia. In this form of aphasia, a patient may claim to know the words for objects but have difficulty in recalling them (Geschwind 1965, Damasio and Damasio 1986, Murdoch, et al. 1986). Varying language deficits, then, may occur from damage to specific areas and also to the circuitry connecting them. And the deficits of aphasic persons are not just limited to language but also extend to motor control and general cognition (see Stuss and Benson 1986).

Recent research has demonstrated that the circuitry involved is much more widespread than the earlier somewhat simpler modular conception anticipated. In other words the formerly designated speech areas have been shown recently to be less specialized than once thought. And these areas are not only interconnected to themselves but also widely connected by neural circuitry to other areas of the cerebral cortex, to the various structures of the

[111]See the model proposed by Schiller (1985) which shows schematically the neuropathways believed to underlie visual-motor control in primates.

limbic system, and into the pathways of the basal ganglia, the midbrain, and brainstem.[112]

The brain seems to be best characterized by occasional pockets of specifically dedicated circuitry (modularity) widely interconnected by extensive neural circuitry throughout the higher cognitive cortex and down through the affectional and life-preserving structures of the mammalian and protoreptilian complexes. All these function together to allow the learning of language and the expression of our thoughts and emotions in social communication.

A VARIETY OF APPROACHES

Researchers approach the study of language from several different perspectives. They define their subject matter varyingly,[113] frame their research objectives differently, and use different methodologies and techniques.

Among these varieties of researchers, we find theoretical linguists who examine the output of language performance and comprehension in the form of spoken or written language, cognitive neuropsychologists who use brain-monitoring techniques, and medical researchers who study the effects of lesions or damage to the neurological equipment of language.

SYNTACTICAL LINGUISTS AND THE SEARCH FOR MODULARITY

Syntactical linguists, for example, analyze the structure of utterances or phrases and compare them across languages looking for common or universal patterns. Some work primarily with one language, some comparatively, over the spectrum of multiple languages. Syntactical linguists tend to focus strictly on what they call "linguistic" factors. They often exclude subjective values, emotions, thought, and related motor activities.

Such linguists may speculate that features or patterns of language are innate or learned, but they do not perform investigations into neural circuits or brain structure. They generally make inferences about such structure

[112]In arguing against the Chomskian notion of language as a discrete, modular device which affected cognition only in specialized domains like grammar and syntax, Deacon has argued that the co-evolution of brain and language restructured the brain and its cognitive functions at a global level. The prominent enlargement of the prefrontal lobes and the shifts in connection patterns, co-evolving with language, gave the frontal lobes a greater role in many neural processes not necessarily related to language (see Deacon 1997: 416-417).

[113]Benson, a professor of neurology at Boston University School of Medicine and a leading researcher in aphasia, has suggested that much of the controversy and difficulty surrounding the diagnosis of aphasia rests with the definition of language itself (1979: 1). It is significant whether the definition of language is limited strictly to basic verbal skills, or whether it is extended to include nonverbal skills, cognition, or emotional components.

from their observations of language performance. Those who speculate from observed language performance tend to think in terms of separate and dedicated language processing modules.

There are probably two reasons for this inclination toward modularity in this group of researchers.

Firstly, if one wants to isolate linguistic factors from all others, one is predisposed to think of modules. Modules facilitate isolation. You tend to see what you're seeking and downplay seemingly extraneous data and issues.

Secondly, many of these linguists work with artificial intelligence and computer languages. Machines such as computers are designed to be modular for purposes of ease and convenience of assembly and repair. In a man-made syntactic processor if syntax goes awry, you unplug the syntax module and plug in another. The language structures of our brain, however, probably didn't evolve so neatly and conveniently. The speculations of such syntactical linguists about the brain structures or the neural architecture of language may or may not bear any relation to the actual brain structure or neural circuitry.

THE COGNITIVE NEUROPSYCHOLOGICAL APPROACH AND THE TENDENCY TO A WIDER DISTRIBUTION

Other researchers, coming from a cognitive neuroscience or psycholinguistic perspective tend to focus, as their name implies, on only the cognitive aspects of language. They may use various techniques to gather data on relevant brain or neural activity associated with language.

In recent years, since about 1980, a good deal of research has centered on monitoring the brain's neuroelectrical responses to and during language activity.[114] This is done by electro-encephalogram (EEG) with electrodes placed at specified external points on the surface of the scalp to record the brain's electrical activity. Such EEG research has indicated language processing and responses taking place in the parietal (top, middle), temporal (side) and occipital (rear) lobes of both the left and right hemispheres (see Neville, et al. 1991).

The rather wide distribution of the area of responses and the involvement of both hemispheres, indicated by this technique has prompted one reviewer to comment that such techniques are difficult to use for localization purposes because of the uncertainties involved in identifying the point sources of the externally, scalp-detected electric fields (Caplan 1995: 874).

[114]Using an EEG, Kutas and Hilyard achieved a correlation of an abstract linguistic response with a neurophysiological measure of neural activity in 1980. The waveforms generated were named event-related potentials (ERPs). The work of Kutas and Hilyard and subsequent investigators showed a clear and specific response to semantic deviance.

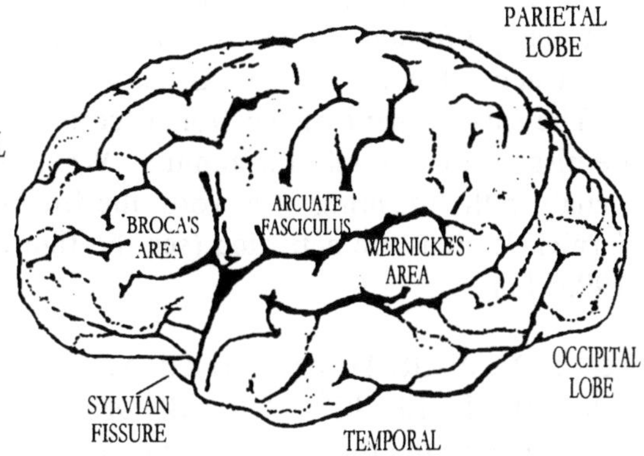

Figure 21-1. A view of the brain from the left hemisphere showing the major lobes, the sylvian fissue, along with the traditional language areas of Broca and Wernicke, connected by a fiber bundle called the arcuate fasciculus. Although some areas are specialized for function, the production of language is now viewed as involving widely distributed areas of the brain, including cortical and subcortical regions.

As cognitive scientist and linguist, Steven Pinker, has observed, research on the neural basis of language is in a state of flux. The former textbook model showing dedicated brain centers for language, such as Broca's area for speech production and Wernicke's for comprehension is pretty much rejected by all neurolinguistic researchers. Pinker, nevertheless, sees a virtual consensus among such researchers that language is concentrated in the perisylvian areas of the left hemisphere which bridge the rear of the frontal cortex with portions of the parietal and temporal lobes. This rather wide area could perhaps be extended to include almost the entire left hemisphere.[115]

[115]Pinker (1995: 852, also 1994: 313). Compare the earlier work of Benson (1979: especially 75-77). A unique study of deaf persons, users of ASL (American Sign Language), who had single strokes in the left or right cerebral hemispheres, revealed that those with left hemisphere damage had language disorders similar to the aphasic conditions observed in spoken language. On the other hand, those with right hemisphere lesions did not show language problems. This led the researchers to conclude that the left hemisphere may be innately predisposed to support the central components of language (Poizner, et al. 1987: 212). On the other hand there seems to be a critical period in the development of children when language can also center in the right hemisphere (Johnson and Newport 1993). In addition some recent research has indicated that the learning of a language may have a modifying effect upon the neurological structure of the brain beyond the mere fact of the modifications associated with encoding. The learning of different languages may program the brain's operating system in differing ways (see Niwa 1989, also Sacks 1989). There are also data from Japanese studies that the Japanese language produces hemispherical shifts (differing

Although not generally considered a language area, research has indicated that the prefrontal cortex is necessary for language encoding and decoding. Studies of regional cerebral blood flow and regional cerebral metabolic rate demonstrate that the prefrontal cortex is always active during language processing, even when other areas (e.g., Broca's and Wernicke's) are not (Ingvar 1983). The involvement of the frontal cortex is also indicated by PET (positron emission tomography) scanning. When a client is asked to think about or analyze the meaning of a word, the frontal cortex as well as Broca's and Wernicke's areas are activated (see the figure and discussion in Kandel, et al. 1995: 17).

MEDICAL RESEARCHERS, LANGUAGE DISORDERS, AND BRAIN LESIONS OR ABNORMALITIES

Medical researchers, on the other hand, have done considerable work on persons suffering from language disorders, most often some form of aphasia. Such disorders involve deficits in speech performance and comprehension and are usually tied to lesions or damage to the brain.[116]

In medical research, the early findings of Broca and Wernicke have been extended to include lesions in the connections between these two speech areas and to language deficits resulting from damage to neural pathways well down into the subcortical limbic centers (e.g., thalamus, amygdala, hippocampus) and the basal ganglia. Findings of these researchers disclose that not just purely linguistic factors are disturbed by such lesions, but that there are also accompanying motor and other behavioral deficits. Such findings indicate that language is not uniquely and completely modular in its structure and function in the brain, but widely distributed and interconnected with other structural and functional circuitry.

---

from Western and other Oriental languages) in some functions and that even speakers of other languages who learn Japanese show signs of this element of hemispherical shift (Tsunoda 1987). The shift in lateralization involves the hearing of human affective sounds such as laughing, crying, plus certain vowel sounds, and the sounds of other animals. The hearing of these sounds is centered in the right hemisphere in westerners. In the Japanese it shifts to the left hemisphere. One possible explanation for this shift is suggested by Levin (1991). The Japanese are culturally more sensitive to emotional context while at the same time inhibited from reacting overtly to it so that the emotional texture should be processed by the left side selectively and analytically to prevent spontaneous, unwanted overt emotional response (Levin 1991: esp. 117-119).

[116]The literature on aphasia is very extensive. Described symptomatically without necessary reference to areas of brain damage, there are roughly seven classifications of aphasia each with its own somewhat distinguishing features: Broca, global, transcortical motor, Wernicke, conduction, anomic, transcortical sensory. For a table of symptoms, see Davis (1993: 19).

THE RELATION OF NONLINGUISTIC FACTORS TO LANGUAGE: PERCEPTION, LEARNING, MEMORY, AND EMOTION

Extending the field further than some theoretical syntactical linguists would care to go, there has been significant relevant research on the necessary relation of perception, learning, memory, and emotion to language production and comprehension. Such research on these related, but essential factors in language takes us even more widely into the brain circuitry. Research reveals increasingly the role of the subcortical brain areas in language. These are the areas also associated with our affectional/ emotional programming.

Study on memory formation, for example, has focused on the hippocampus. Damage to the hippocampus, a structure of the limbic system, has long been known to effect memory. Since the discovery in 1966 of the phenomenon of long-term potentiation (LTP), and the subsequent reporting in 1974 of LTP in synaptic transmission, the phenomenon has been studied especially in the hippocampus. LTP is a form of neural plasticity related to both pre- and postsynaptic changes associated with patterns of electrical activity. Because of its lasting effects (among other reasons), LTP has become the model of choice for an explanation of biological information storage at the cellular level.

An exciting, multidisciplinary field has grown up around seeking the causal relationships between the LTP phenomenon and learning and memory. This is significant for language because you can't have learning, to include the learning of words for language, without memory.[117]

The lexical items, of language, our vocabulary, then, are dependent on processing through the hippocampus, amygdala,[118] as well as the structures of the thalamocingulate division which as seen earlier are involved in nursing, maternal care and family behaviors.[119] These limbic structures, perhaps most of them, add emotion or significance to the items before they are assigned for storage in long term memory in the higher cortex.

[117]For a review of the research on LTP, see the compilation of articles in Baudry and Davis (1991).

[118]For a compilation of recent research on the amygdala, a previously neglected part of the limbic system adjacent to the hippocampus, see Aggleton (1992); also LeDoux (1996). Research has shown the amygdala's involvement in cognition (Halgren, 1992), emotional behavior (LeDoux, 1992), and social affective behavior (Kling and Brothers, 1992).

[119]The thalamus is definitely involved in language function, although as in the case of other limbic structures, the extent of involvement is not yet well understood. For example, the syndrome of thalamic aphasia results from lesions in that structure. See Benson (1979: 93-97).

Bernard Donovan, a physiologist at the Institute of Psychiatry in London, in his report on hormones and the mind, indicates that the evidence on learning and memory points to wide-ranging interaction among the perceptual apparatuses, the limbic system, and the cerebral cortex. Donovan's suggested schema for learning and memory, which accords well with current research, provides that information is sent from the five senses via the cerebral cortex to the limbic system, where if it is deemed significant by the various limbic structures the emotional connotations are added, and it is projected back to the association cortex for storage.[120]

That is, words or experiences, are passed through these earlier mammalian and protoreptilian structures, evaluated for signficance, and given their affective emotional charge before being sent to the cortex for final storage in long-term memory. And, of course, they tend to evoke this charge when recalled. This accords well with Paul MacLean's statement concerning the role of the limbic system in learning and memory that objects do not exist for us unless "imbued with an affective feeling, no matter how slight." (MacLean 1990: 578).

Although some learning seems to occur sometimes gratuitiously simply because the mechanisms are there (e.g., nonsense rhymes, trivia) or because they are incidental or contextual tag-alongs with learning experiences of significance (e.g., see Kosslyn 1995: 370; McClelland and Rumelhart 1986), essential memory, the most long-lasting and significant memory is that memory which is keyed ***somatically*** to the our survival and affectional mechanisms at both very specific (pain, hunger, nursing, physical contact comfort, sex) and highly generalized levels (ego and empathy).

On a personal note, I remember very specifically and vividly almost cutting my thumb off with a broken bottle at age three. I also remember the general process of studying intensely to learn the material necessary to get a Ph.D., which learning and effort were keyed significantly to my ego and also

[120]Donovan (1988: 200). See also Bachevalier and Mishkin (1984) who found two learning systems in infant monkeys which developed at different times: a habituation system involving the phylogenetically earlier and ontogenetically developmentally earlier lower brain and cerebellar structures, and a cognitive memory system dependent on the maturation of the limbic (esp. hippocampic) circuitry. Altman and Bayer (1997: 749) in their comprehensive report on the development of the cerebellar system speculate that the ease with which the human child learns complex motor, to include language skills, is because of a critical period in cerebellar plasticity. Such learning becomes hardwired. In contrast, the adult depends on motor and language learning mediated by the cerebral cortex, which require conscious effort, is softwired and lacks the automatic, resistant quality of the earlier cerebellar learning. Nevertheless, the two systems interact through the transcerebellar loop and their contributions become interwoven in any specific instance of motor or language performance. Compare Diamond (1991) whose research also shows the importance of prefrontal cortex maturation. Also Janowsky (1993).

my empathy (to be able to better provide for my growing family and, perhaps, even to contribute to humanity). The subjective semantic component of language, then, is tied to the second algorithmic somatic matrix, the tug and pull of ego and empathy, as mediated through the source mammalian and protoreptilian programs, and subjected to recall from storage in the higher cortex where the words and their semantic charge are evaluated "rationally" as they are being put into the language of thought and social communication.

BIPED LOCOMOTION AND THE VOCAL APPARATUS: THE SPEECH CONNECTION

The evolution of language is not just a matter of changes in the size and organization of the brain and related neurological equipment. With the emphasis on the brain, the concentration on language centers or modules in the cortex, and more recently the subcortical centers, other physiological changes necessary to the evolution and production of language tend to get overlooked. Among these are the structures of bipedal locomotion, which freed up not only the forelimbs but also the jaw and related structures. Bipedal locomotion freed up the forelimbs to develop into our dexterous hands with opposing thumb and fingers so that we could use and develop tools. Bipedal locomotion also freed up the mouth and jaw from doing the multiple tasks of not only eating, but fighting, capturing, and transporting. The freeing up of both the hands and the jaws was undoubtedly prerequisite to the evolution of higher social skill, cognitive skills and the likely parallel development of spoken language.

The linguistic psychologist, Philip Lieberman, has explored the extensive evolutionary modifications which occurred in the transition from the characteristic mammalian vocal tract to that of homo sapiens. Necessarily accompanying the significant changes in physical structure of the nasal passage, mouth, larynx, and pharynx was, of course, the compounding of neural circuitry to support the new structures and functions (Lieberman 1991, 1998). This circuitry runs from the muscle structure of the vocal apparatus up through the brainstem, cerebellum, basal ganglia, limbic structures, and into the cerebral cortex itself. Such circuitry is necessary for the extraordinarily fine motor control required for speech.

It is worth noting that the human basal ganglia, part of the earlier reptilian complex through which much motor circuitry of the body runs, when adjusted for weight differences, is approximately fourteen times the mass of an insectivore, a primitive mammal from whose ancestors primates likely evolved (Parent 1986:125). The additional circuitry required to support speech in humans undoubtedly contributes significantly to this difference.

THE FULLY INTEGRATED MOTOR-MOUTH

When it comes down to it, almost the entire brain is involved in the language function.[121] Our higher cognitive areas are wired in, or we would not be able to express our thoughts even as well as we do. Our visual cortex is wired in, or we wouldn't be able to describe what we see. Our auditory areas are wired in, or we couldn't learn or repeat the sounds we use.

It is also the case with our other senses. We can talk about what we touch, what we taste, and what we smell because these circuits, too, are tied into the language-making network. Even the cerebellum, located at the rear base of the cerebral hemispheres, and intimately involved in coordinating motor activity and balance, has been shown to be involved in the language function...the motor control of speech (Altman and Bayer 1997, MacLean 1990: 548-549, also Dow 1974, and Leiner, et al. 1986).

THE WIDELY DISTRIBUTED, HIGHLY GENERALIZED NATURE OF LANGUAGE

The mechanisms of language are clearly highly generalized. Focus on specific areas and functions in research tends to obscure this obvious generalization. The evolution and development of language may or may not have depended on the evolution of the speaking apparatus, but language is now unquestionably a generalized motor activity, not dependent on the vocal apparatus at the present stage of evolution and development.[122]

Whatever the evolutionary steps (and in the absence of convincing paleontological evidence, we can debate those interminably: i.e., big brain first, language second; language first, big brain second; evolution of vocal apparatus, language second), now that we've got it, language can be expressed through any and every motor channel.[123]

We don't, for instance, write with our speech organs. We write with our hands. We could, however, if we lost our hands, write with our tongues, by licking and smearing. We could, also (and this is easy to verify), write with

[121] Or, as Deacon suggests, the coevolution of language and brain restructured the brain at a global level (1997: 416-417).

[122] See Bellugi, et al. 1989). These researchers report that users of ASL (American Sign Language), a visual-spacial mode of language, develop the equivalent syntax and lexical sophistication of speech-based language, also specialized in the favored left hemisphere.

[123] Petitto (1997) presents evidence and argues for equipotentiality of spoken and signed language modalities and posits a genetic, nonmodality-specific or amodal cognitive structure-mechanism underlying them. She reports that deaf and hearing children acquire signed and spoken language in virtually identical developmental ways despite the significant differences in the acoustical and visual modalities. From birth though age 5, both speaking and signing children exhibit identical stages of babbling in both modalities.

an instrument held between our teeth, lips, toes, arm and leg stubs. We can also communicate language by tapping with any part of our body, by the well-known example of signing used by the hearing impaired, by nodding our head, or blinking our eyes, or even banging our head against the wall in morse code.[124]

We can do language through any motor medium available to us whatever.

THE GENERALIZED AND FLEXIBLE CHARACTER OF CORTICAL CIRCUITRY MECHANISMS

Although the neocortex (highly developed in mammals), can be divided into a number of specialized areas based on structure and function, there is also increasing evidence of structural and functional consistencies from one area to the other. In other words, the cortical mechanisms underlying the varied functions and structures of our higher cognitive circuitry seem less specific and more generalized and flexible than previously thought. This has led a number of neuroscientists to propose that primary cortical areas, despite differences in function and structure, share a common organizational scheme (Eccles 1984, Powell 1981, Mountcastle 1978, Lorente de No 1949). Cognitive neuroscientist Dennis O'Leary, has reviewed some results of research on re-routing of connections and transplantation of tissues in the neocortex of mammals. For example, some experiments re-routed visual input to primary auditory or somatosensory areas of the cortex of hamsters. As a result, visual information not only found its way to the new target areas but the auditory and somatosensory cortices took on actual visual functions.

Research on monkeys has shown that tissue from one area of the cortex transplanted to a second area takes on the functional properties of the second area.

Such findings strongly indicate that the cortex is composed of building blocks which are to some degree equipotential and can be made to serve various functions. Such functions may be determined by location in the cortex, connections to subcortical structures (e.g., limbic structures like the thalamus) and other events that occur in the process of development (O'Leary 1989). True, there are areas that focus specific functions, like Broca's and Wernicke's. But these depend on connections throughout,

[124]Gomez (1998), in discussing the ostensive/inferential communication system of chimpanzees and gorillas, notes that virtually any behavior can be used to communicate; e.g., movement of eyebrows, adopting a posture, looking or walking in a direction (81-82). He suggests that the ostensive/inferential dimension reflects a continuity between the great apes and humans and that such a system may be an evolutionary precursor of syntax. He further suggests that syntax and semantics may be considered tools primarily serving this more powerful system (84). See Taylor (1997, esp. 241-260) for an argument against the existence of a language gap between humans and the higher primates.

between each other, among other cortical areas, and down through the subcortical areas of the early mammalian and protoreptilian complexes. And the left hemispherical lateralization of language, although highly favored, is not absolute. The right hemisphere can host the language function if the left is damaged early in development (see Neville 1993: 435). The right hemisphere can also be specialized for language even if the left is not damaged. In fact, recent research has shown that the right hemisphere is much more involved in language processing that previously supposed (Beeman and Chiarello 1998).

Furthermore, although the left hemisphere is specialized for language in about 90% of humans, the right hemisphere is the language hemisphere in about 35% of left-handed persons and in about 5% of right-handed persons (McManus and Bryden 1993: 680, 692). The specialization seems to depend upon the achievement of competence in the grammar of a language independent of the modality; that is, whether the language is expressed in sound or visually as in sign language.[125]

SUMMARY

We are a hierarchically integrated, but flexible and, more or less, highly modifiable, motorized bundle of interactive programs...interactive among each other, and interactive with the physical environment and with the social environment.

Despite the complexity, the master algorithm, the primary algorithm, peers through as the guiding framework and process. The billions of neurons are intricately interconnected into hierarchical networks supporting subroutines that are further interconnected and integrated to produce a unified language function and a conscious self...a self that experiences the neural function as well as applies it volitionally as a unitary, highly generalized, synthetic, and self-conscious entity.

This hierarchical complexity can be represented at varying levels by the algorithmic structure of the binary or multi-node hierarchical tree. At one level neurons and/or their dendritic webs and connections, may be seen as nodes in a neural network. At another level, groups of neural nets form single nodes at higher levels of organization, and so on. And we still do not know with any certainty what hierarchical processing goes on within the single neuron itself, although the indications are that it may be considerable.

The detail and variety of all life evolved by the complexification of primal protein nucleic-acid synthesis and its related metabolic activity. The

[125]This conclusion is supported by the research reported in Bellugi, et al. 1989 and Neville 1993.

interconnected, hierarchical architecture of the neural networks is guided by and expresses the fundamental algorithm of that synthetic process...the ordering, incorporating, synthesizing algorithm that establishes, and still functions to mediate, our subject-object relationship with the environment.

PART FIVE

FURTHER THOUGHTS AND CONCLUSION

Chapter 22

FURTHER THOUGHTS ON THE NEURAL ORGANIZATION OF THE PRIMARY ALGORITHM

Neurons are generally involved in the integration of multiple synaptic inputs....

(J.T. Coyle and S. E. Hyman, Psychiatrists, 1999: 4).

The primary algorithm is the incorporative, structuring algorithm of the protein-nucleic acid synthesizing loop. The synthesizing algorithm involves a symbiotic, interactive relationship between proteins (to include the amino acid building blocks) and the nucleic acids (DNA, RNA). That is, proteins and nucleic acids mutually and reciprocally build each other (e.g., see Margulis and Sagan 1995: 50). In molecular neurobiology both proteins and nucleic acids are referred to as information molecules (Smith 1996: ch. 2). The primary algorithm, then, as an incorporative, structuring algorithm is specifically an information processing algorithm. It is replicated as the primary algorithm of our cognitive structure. It is a fractal-like,[126] cybernetic algorithm expressed dynamically in neural flow and information processing and more statically, in neural structure and in storage of knowledge and memory.

Previous chapters of this book have traced the derivation of the primary algorithm from physics and biology and demonstrated its ubiquitous expression across the spectrum of human thought, language, experience, and

[126] As Lowenstein notes, the information repository grew more by building up the number of loops rather than by increasing the content within the loops (1999: 109). This is in the nature of fractals which generate the same pattern repeatedly at different levels to achieve great complexity out of essential simplicity.

behavior. This chapter, will look at the question of how this behavioral-experiential algorithm is generated by, or actually maps on to, the physical flow and structure of our neural system. Our knowledge of the brain's structure, although still limited, is probably sufficiently known to give us a general idea of what this generative neural mapping process could look like.

That the mapping process occurs is without doubt. We ***do*** learn and we ***can*** recover our memories, and we ***do*** reprocess our memories, our learning, into new forms and ***apply*** them in coping with the everyday activities of life and survival. This is all done in the primary algorithmic framework and process.

COGNITIVE NEUROSCIENCE AS THE PURSUIT OF THE ALGORITHM

The vocabulary of the primary algorithm is expressed throughout the writing of cognitive neuroscience. For example, take the following quotes from various writers:

> The second type of information ***incorporated*** (emphasis mine) from the environment is referred as experience-dependent...this refers to information ***absorbed*** (emphasis mine) from the environment...unique to the individual. (Johnson 1993: 286).

> The extended period of infancy reflects the importance of ***incorporating*** (emphasis mine) enormous amounts of information into the brain. (Greenough, et al. 1993: 290).

> As the brain ***incorporates*** (emphasis mine) dispositional representations of interactions with entities and scenes relevant for innate regulation, it increases the chances of ***including***(emphasis mine) entities and scenes that may or may not be directly relevant to survival. (Damasio 1994: 117).

> Alternatively, the organism gains in form by ***absorbing*** (emphasis mine) the ***structure and patterning*** (emphasis mine) of its physical or social environment through its interactions with that environment. (Thelen 1989: 77).

> Once this system (visual) is operating, infants can learn where to attend in ***order*** to ***absorb*** (emphasis mine) the information their culture values. (Posner and Raichle, 1994: 203).

> Thus at the behavioral and the neural levels, recognition involves both ***specificity*** and ***generalization*** (emphasis mine). (Horn 1993: 494).

> The ***discrete combinatorial*** system (emphasis mine) called 'grammar' makes human language infinite (there is no limit to the number of complex words or sentences in a language), digital (this infinity is achieved by ***rearranging discrete elements*** in particular ***orders***....) (emphasis mine). (Pinker, 1994: 334).

> Some connections are stabilized by being ***incorporated*** (emphasis mine) into functioning systems either through the establishment of neural circuits or through afferent activity transmitted from sense organs...Cerebral ***metabolic*** rate (emphasis mine) is closely linked to the synaptic activity of cortical neurons....(Huttenlocher, 1993: 120-121).

Molecular biologist, Werner Lowenstein tells us about the fundamental nature of biological communication in the primary algorithmic terms matching the dialectic of logic and thought,

> Informational ***convergence*** (emphasis mine) and ***divergence*** (emphasis mine) are the hallmarks of biological communication. (Lowenstein 1999:308).

Convergence is the assembling of parts into wholes (induction or synthetics); divergence is the disassembling of wholes into parts (deduction or analytics). The progression from convergence to divergence to convergence, etc., characteristic of neural connectivity, also describes the flowing process of the philosophical dialectic.

The primary algorithm and its comparability with or derivation from protein-nucleic metabolism is expressed clearly and straightforwardly from an ethological perspective by Eric Lenneberg of Harvard Medical School when he writes that the individual human being can only function linguistically if he can

> ...synthesize...the entire language mechanism out of the raw materials available to him. The raw material is of no use unless it can be broken down as food proteins are broken down into amino acids, and built up again into the pattern of his in-dwelling latent structure. (Lenneberg 1967: 378).

In fact, the entire project of cognitive neuroscience concerns the study of the ***discrimination, incorporation, ordering, synthesizing***, of information from the environment, the processing of such coded information as necessary into new ***syntheses*** and ***generalizations*** at all levels, and the workings of the underlying neural structures and processes that make all this algorithmic activity possible. It might be worth reminding ourselves again that the term ***incorporation***, which means basically to take something into and make it a part of the body, is appropriate terminology for the enterprise. We now have clear evidence that such incorporation of learning and experience does not belong to something insubstantial and abstract called mind. It involves actual physiological modifications to our bodies...to the protein synthetic process and its associated metabolism.

NUCLEIC ACID INFORMATION PROCESSING VS. PROTEIN INFORMATION PROCESSING

In the biomolecular world there are two important and distinct categories or realms of information processing. First, there is the information processing of the ancestral genetic code which translates the inherited DNA code via RNA into the essentially temporary, phenotypic protein structure of the individual...from simple one-celled organisms right up to complex vertebrates. Second, there is the information-processing done by the proteins themselves. These proteins not only implement genetic information to maintain their structure and function, but also process external signals and compute the activities that control or effect behavioral exchanges with the environment.

Genetic information, coded and transcripted by the nucleic acids, is relatively permanent...at least for the individual organism. It is fixed in the genetic informational matrix by evolutionary processes. Taking DNA as the starting point, there are two streams of information flow going forth. The main flow of genetic information is one way...from DNA to RNA to the creation of the proteins necessary to the structure and living functions of the organism. As noted biologist, Francis Crick put it: *DNA makes RNA and RNA makes protein.* Although this has been called the central dogma of molecular biology, there are some exceptions when information flows from RNA to DNA (Smith 1996: 46).[127] The second information flow starting from DNA replicates the genetic informational matrix (DNA) itself, with the assistance of protein catalysts. The protein-nucleic acid building processes do, in fact, in accord with the synthesizing loop, flow circularly...that is, nucleic acids build proteins and proteins, in turn, function reciprocally to catalyze the replication of the DNA matrix.

Nevertheless, the flow of temporally or phenotypically modified protein information does not flow back to modify the genetic code. That is, environmentally interactive behavioral information processing by proteins does ***not*** flow back through RNA to DNA...at least ***not*** as a permanent change in coded DNA information storage. A reversal of the protein-mediated temporal behavioral flow into the genome would constitute an instance of Lamarckism, or the transmission of acquired rather than inherited characteristics. There has yet been no convincing instance of Lamarckism in biology to date (see Smith 1996: 46-47).

So we can safely say: in the main informational stream, the genes (DNA), through the mediation of the sister nucleic acid RNA, guide the processing of hereditary information into proteins; proteins, then, additionally and principally execute the imperatives of genetic information matrix, based upon which they process behavioral informational interactions with the environment. And, of course, as the proteins are modified in the environmental exchange, learning occurs.

The informational modifications of the proteins, resulting from the sensory and behavioral exchanges with the environment that constitute learning, are ***not*** passed through to the ancestral coded information of the

[127] See also the discussion of possible reverse transcription from RNA to DNA in Lowenstein (1999: 149-154).

DNA genome. The reverse transcription is blocked by molecular configurational barriers. The three dimensional, folded configuration of proteins does not allow the one-to-one complementary matching to the more linear nucleic acids which would be necessary for unit by unit transcription (see the discussion in Lowenstein 1999: 108-117).

In unicellular organisms lacking neurons and a nervous system, protein-based circuits, function to control behavior. Such protein circuits process information, do computational tasks, and store information. Their information storage capability has been compared to ***random access (RAM)***[128] ***memory*** in contrast to the permanent genetic information of DNA (Bray 1995). Proteins, then, are the fundamental units of both intra- and intercellular computation from a neural network viewpoint. Neurons, those cells specialized for fast, relatively long-distance communication throughout a complex multicellar organism, represent a subsequent evolutionary cellular development. They are cells specialized (like cells of other bodily organs) to perform and elaborate this already present elementary protein function of behavioral information processing. In unicellular animals, like bacteria, then, behavioral, environmental experience, is learned and memory stored by physical incorporation into the protein circuitry of the cell. In higher organisms like ourselves, research shows that the laying down of longterm memory requires the synthesis of new proteins (see Dubnau and Tully 1998, Smith 1996: 419-443, Kandel, et al. 1995 and the summary in Macphail 1993: 59-61). Such learning and experience then become literally and actually a part of the structure of the body...truly incorporated into that evolved and modifiable three dimensional protein structure which constitutes our bodies. The mind/body connection…the somatic basis of cognition… remains firmly in tact.

The incorporative algorithmic loop surely expresses the varying modifications in neuronal protein structure and wiring to produce its differentiated structuring and functional effects for each sensory, memory, thought, and language-making modality. Lowenstein, for instance, writes that organisms evolve not by adding content to the informational loops, but by adding loops upon loops in a seemingly endless accumulation of interconnected complexity. Recognition of this may provide the middle ground between the strict modularists like Chomsky and Pinker and those like Lieberman who see such distinct modularity as biologically improbable.

As we have seen in the previous chapter, the rather general purpose and consistent cortical neural structure can be made to perform other functions

[128]RAM in this usage would, of course, not be exactly analogous to the computer usage, but would have to include implicit and explicit memory as well as short term and long term memory. In other words the protein-based storage would accommodate all *phenotypically* acquired memory. DNA/RNA code and transcribe *genotypically* acquired memory.

by either transplanting of tissue or altering efferent and afferent (outgoing and incoming) pathways. In other words, ***it would not take monumental evolutionary cortical changes to produce language from thought or vice-versa.*** With such rather versatile neural cortical mechanisms, simply having more of them freed up from basic survival, sensory, and motor routines might serve to move from simple (implicit) forms of learning and thought to more complex forms as the brain's somewhat general purpose raw materials of the cerebral cortex, are rewired more intricately, not only by evolution, but also by experience and usage.

THE PRIMARY ALGORITHM AND THE NEURAL NET

Advances in molecular neurobiology are changing our traditional conception of neural nets. In recent years the informed estimates of the number of neurons in the human brain have risen dramatically from figures in the low billions to a current generally accepted figure of 100 billion (e.g., Kolb 1993: 340, Huttenlocher 1993: 116-117). The figure will probably grow larger in the future.[129] It has further been estimated that the average neuron makes up to 15,000 points of connection with other neurons (see Cowan 1979) with some neurons in the visual cortex having up to 30,000 synaptic inputs (Greenough, et al. 1993: 295) and in the cerebellum up to 150,000 contacts (Kandel 1995: 27).

By any standard that is a lot of connectivity! And graphing in any detail the connectivity of a few neurons on the standard algorithmic binary tree would bring this complexity to one's attention dramatically. Despite the rather substantial work already done in developing artificial neural nets, such nets are still at best very far from any approximation of the brain's vastly complex information storage and processing structure and capabilities. In fact such nets are still judged to be simplistic and naive compared to those of the brain (Changeux and Dehaene 1989, Elman 1995: 512).

IS THE ALGORITHM WE EXPERIENCE LIKE THE ALGORITHMIC STRUCTURE WE SEE?

Physically viewing the organization of neural networks, we see they are characterized by connectivity and hierarchy. At one level the neurons themselves, with their associated dendritic trees, are nodes interconnected multidirectionally and hierarchically into nets...sophisticated and hierarchical tree-nets.[130]

[129] In fact, Lowenstein gives an estimate of 1000 billion (1999: 307).

[130] For a hierarchical development model from a psychoanalytic perspective, see Gedo and Goldberg (1973). Levin (1991) gives an updated treatment of Gedo's work with recent research and discussion.

It appears further that the phenomenon of LTP (long-term potentiation), which apparently supports learning in the hippocampus, necessitates the firing in both the presynaptic and postsynaptic neurons. Kandel, et al., point out that this provides the first direct evidence for the rule proposed by neuropsychologist Donald Hebb of McGill University in 1949 (Kandel 1995: 27). Hebb's rule stated:

> When an axon of cell A....excites cell B and repeatedly or persistently takes part in firing it, some growth process or metabolic change takes place in one or both cells so that A's efficiency as one of the cells firing B is increased. (Hebb 1949).

But the empirical evidence that the cognitive expression of the incorporative algorithm goes right down even into the cellular level of protein synthesis is increasingly apparent. Research, for instance, has shown that in the hippocampus of rats the environment is mapped by ***place cells***. These place cells are single neurons which fire when the rat is at a specific place in the environment and only at that place (see the summary article by Stevens 1996; also Jung 1995; O'Keefe and Dostrovski 1971). These various place cells are organized into a cognitive map that represents the entirety of the particular environment, say a countertop. Such place cells have also recently been found in the hippocampus of primates (Macaca fuscata) (Ono, et al. 1995). Single neuronal responses have also been reported for words, faces, and particular objects (Perrett, et al. 1987, 1995; Heit, et al. 1988; also see Changeux and Dehaene 1989).

Since place cells which map adjacent spots in the environment are not necessarily adjacent to each other in the hippocampus, the neural map is not a one-to-one topological map but a cognitive map of some sort that is not yet fully understood. The fact that the place cells cover an area and not a point (corresponding to a bit of information), indicates that the neuron itself is a processor that assembles bits into patterns rather than functioning only as single point node.

Recent research on the single cell *E.coli* bacterium supports the probability that single neurons can detect such spaciotemporal patterns. The single-celled *E.coli* apparently has information processing capabilities that were traditionally thought to require the combined working of many nerve cells or neurons. Horace Barlow of the physiological laboratory at the University of Cambridge estimates that *E.coli* is about the size of a single spine of a cortical pyramidal nerve cell which has about 5000 such spines. The total volume of the pyramidal neuron equates to roughly a quarter million *E.coli*. Barlow suggests these facts may revolutionize thinking about computation in the brain; since the intracellular level of computational processing has previously been largely ignored (Barlow 1998).

In fact, the new discipline of molecular neurobiology has clearly repudiated the earlier, but still lingering conception, of the neuron as a simple on/off relay, that either fires or does not to communicate its single bit of information. The neuron, like any single cell with its DNA/RNA and protein computational properties, is, in reality, a tiny computer in its own right, a microprocessor (Smith 1996: 9-15). Although, like any other cell, it seems to process predominantly genetic information through its DNA, RNA, and protein macromolecules, as noted earlier, it indisputably also processes experiential information by protein-based circuits through its connections with its surrounding environment. And it is yet unclear exactly what the neuron is communicating, when the neuron fires in its one way, polarized axonal direction. The action potential may be communicating or transmitting a hologram of the relevant stored information in its dendritic network, as suggested by Pribram (1991, 1998). Or it may be transmitting information by temporal coding which would permit multiplexing of information in the time domain. This would suggest a broadcast model of information transmission in contrast with the traditional all or nothing spike tied to specific connectivity (Cariani 1994). Or it may not be "transmitting" anything in the sense of sending it somewhere else. It may be simply ***coming on line***...becoming even part of the dynamic core of consciousness postulated by Tononi and Edelman (1998).[131] After all, we ***are*** the neural network; that is, we are our brain...or more Pogoish, ***our brain is us***. We don't necessarily have to access it like an outsider. As philosopher of science Daniel Dennett observes, we automatically know about these things going on in the body because, "...if you didn't, it wouldn't be your body!" (Dennett 1998: 293). The idea that we "access" parts of the brain's networking apparently comes not only from Cartesian dualism but also the positivist assumption of an outside objective observer (or even an internal homunculus) which so commonly accompanies the externalized scientific perspective.

Kandel, et al. have recently raised the question of a molecular alphabet for learning. They point out that "...synaptic changes can be associative without depending on the complex features of the neural network." This fact further strengthens the proposition that the associative activity that contributes to explicit and implicit learning represents a basic cellular process (Kandel, et al. 1995: 685).

This basic cellular process, of course, is a protein synthetic process of ***incorporation*** which creates a ***modification in cellular structure or the***

[131] The idea of ***coming on line***, as opposed to going somewhere, is consistent with the growing idea of ***integration by time***; that is, our sense of all things coming together somewhere in the brain is probably an illusion based on the synchronous neural activity of perhaps numerous separate brain regions (e.g., see Damasio 1994: 84, 95, 276-277; Llinas 1993; Zeki 1993; Edelman 1987; also Calvin 1996).

associated metabolism which supports what we call information or learning. At the next level, these basic cellular incorporative learning mechanisms are "imbedded in neural circuitry with considerable additional computational power, which can add substantial complexity to these elementary (cellular) mechanisms" (Kandel, et al. 1995: 685).

That is, at a higher level of hierarchical organization, the neural nets may act as nodes and are interconnected hierarchically into increasingly higher level nets and so on, with concommitantly increasing computational power. Connectivity and hierarchy constitute the relationship among nodes. Electric potentials and neurotransmitters are the energizers or innervators that run across the synapses and up and down the dendritic-axonal wiring and facilitate the functional job of computation and information processing within the wiring structure.

The new connectionist models used by neurobiologists working in artificial intelligence have abandoned the former serial processing concepts and turned to parallel and distributed components. This means that there are multiple pathways and neural groupings processing the same information.

Individual neurons, as the basic computational and signaling units, may or may not transmit large amounts of information, but they can undoubtedly participate in important computations because of their extensive and diversified interconnections. Complex mental faculties are not themselves performed by discrete local regions in the brain, but are built up from more elementary operations which tend to be more specifically localized. This diversified and extensive wiring of the neural circuitry is not irrevocably fixed and predetermined but plastic and modifiable. Modifications occur throughout development and, later and more extensively, through lifelong learning. The plasticity in the relationships among the neuronal units allows for the emergence of considerable individuality.[132]

WORKING ONE'S WAY UP AND DOWN THE NEURAL NETS: THE PHYSICAL SUBSTRATE OF INDUCTIVE AND DEDUCTIVE REASONING?

The basic unit of the neural net is the neuron and the basic function of the neuron is clearly inductive. The neuron's molecular machinery inductively sums the largely excitatory inputs from its dendritic synapses and its own internal largely inhibitory regulators into a synthetic go, no-go decision to fire an action potential along its axon to communicate its synthesized information within the neural net (e.g., see Coyle and Hyman 1999: 3-6). Further, recent findings tell us that mental computations seem to involve

[132]Kandel, et al. provide a general discussion of these issues in the introductory chapters of their textbook (1995: 5-40). See also Nestler and Hyman (1999: 61-71).

roughly the same anatomical area whether the area is activated by top-down signals from the higher cortex or by bottom-up signals from lower sensory areas within the neural network(Posner and Raichle 1994: 131-152).

Accordingly a direct isomorphism may be suggested to exist between our inductive and deductive thought processes and the physical structure and function of the brain's neural nets. Take a major synthetic thought--all the parts that would go into the make-up of such concepts as ***universe*** or ***molecule***, ***nation*** or ***society***, ***development*** or ***computer program***. Each of these can be built up (inductively) from many parts or broken down (deductively) into many parts. In each case the parts (bits or clusters of bits) would probably be represented by an interconnected, hierarchically-structured set of neural nets. By an act of conscious or executive choice executed from the frontal cortex, probably from what Posner and Raichle propose as the executive attention network (see also Baars 1997, Harth 1997, LaBerge 1995, Crick, 1994), we can access or bring on line and follow down through the thalamocortical gating circuits into the interconnections of the net and ***analyze*** the constituent parts of the overall net ***deductively*** or else ***synthesize*** them ***inductively***.[133] We may have conscious access, or the capacity to bring on-line, even the inner workings of the cellular structure although such cellular modifications may well constitute what is called implicit (or nondeclarative) learning, and may not be subject to access and conscious recall by the frontal cortex.

Since every act of learning, memory, or even modification of existing learning or memory involves a physiological change in the neural nets or neurons making up our brain,[134] we obviously have the capacity to modify existing connections, or create new ones by ***conscious*** and ***willful***, as well as ***automatic*** and ***fortuitous***, acts of learning, thinking, and behavior.

[133]See Posner and Raichle (1994) for the model of an executive attention network. The anterior (frontal) cingulate gyrus lying in the center of the frontal lobes seems to be the control center. It acts in concert with other frontal areas which hold the representation of information (in the absence of a stimulus event) in working memory while processing, including modification as appropriate, is performed upon the information represented, (see especially 1994: 168-179). Citing the work of Fuster and Bauer (1974) and Goldman- Rakic (1987), Janowsky (1993) notes that the frontal cortical areas have been shown to be critical for the moment-by-moment updating of representations in memory. In addition the frontal cortex is necessary for associating representations with their unique contexts in time and space (see also Janowsky, et al. 1989; and Milner, et al. 1985). See also the research report by Waltz et al. 1999 which indicates involvement of the prefrontal cortex in high level and self-conscious deductive and inductive reasoning tasks.

[134]Greenough, et al. (1993) also support the claim that individual learning (information storage) involves the generation of new synaptic connections even in the adult. The forming of these new synapses is coupled with changes in blood supply and astrocytes (glial cells) that metabolically support the neurons and their synapses. Note especially the update on 319-321.

It is probably better to consider that ***we are the neural network***, therefore that we do not access it like an outsider, but rather activate it by conscious, deliberate effort from our prefrontal executive attentional centers. Of course, the more basic, automatic survival centers can, by being stimulated or aroused, open the thalamocortical gateways and enter the cortex or else come on line.

There are assuredly ultimate limits on our capacity to learn and modify our neural circuitry. Such limits would be imposed by our genes as well as environmental constraints. Some of the limits are even time critical. But as long as we can still think and learn we obviously haven't exceeded the limits. The proof is in the pudding. We can either learn more or not. It is an empirical state that can be measured. Most of us probably never exhaust our capacity until stopped by brain damage, brain disease, or death, not only because we only use a part of our brains, as is so often pointed out in the popular literature, but also because many of the parts in use can be modified and reprogrammed.

THE DILEMMA OF CONSCIOUSNESS

We can see the outline of the physiological substrate (the neural connectivity and hierarchy) of our memory and thought processes rather clearly and it does seem to match our experience of it. Does this mean that we know what consciousness is?[135]

Yes and no. Although we can see increasingly clearly the physiological neural substrate that underlies consciousness or from which consciousness emerges, this observational knowledge still does not tell us what it feels like to be conscious. In confronting the dilemma of consciousness, mechanical analogies, whether switchboard, computer, or moving vehicle analogies, may be helpful, but they can take us only so far in the direction of understanding ourselves. If we think of ourselves, for instance, as a computerized vehicle, knowledge of the parts and discovery of the design of the vehicle may be very helpful for the preservation, maintenance, repair, and to some extent the application and use of the vehicle. Such knowledge and discovery, however, do not combine to tell us about the experience of driving the vehicle.

Our situation, moreover, is not covered well even by the analogy of driver of the vehicle. In the case of our brains, we are not just the operators of the vehicle, ***we are the vehicle itself.*** It is therefore noteworthy that our thought, behavior, as well as the subjective experience of the vehicle's operation

[135]For a recent article on the puzzle of conscious experience, see Chalmers (1997).

matches the structure and process of the same primary bioneurological algorithm.

On the perplexing problem of consciousness, we may eventually be able to identify and fully define the neural substrate. Even this degree of absolute knowledge, however, will still not tell us what the experience of consciousness is...how it feels to be conscious. Monitoring of electrical potentials, images and pictures of neurons and networks of neurons firing away simply cannot capture the subjective experience. For an understanding of the texture, the subjective essence of consciousness, we must consult the vehicle itself. That is we must look to personal accounts of the processes and products of our behavior, thought, and subjective experience... personal accounts of ourselves and others like us. There is presently no other empirical way.

At any rate, the potentially completely definable, observable connectivity and hierarchy of the intra- and intercellular nets, amplified by the specialized communicative neurons and neural networks of our brain are the physical substrate of the cognitive primary algorithm. This physical representation of the algorithm apparently matches our behavior, thought, and the subjective experience of it...from the inductive/deductive dialectical processes of thought up to and including the subjective experiences which we have traditionally described as mystical.

The linkage or connection between neuroscience and our thought, behavior, subjective experience, and social life is established. The cognitive, fractal-like behavioral/experiential algorithm seems to have its counterpart, indeed its actual source, in the neurological structure evolving and growing out of protein-nucleic acid synthesis. That is, the neurological structure, too, appears fractal-like.

In other words, we have points of reduction, or better, points of connection or links between the behavior, experience, and the neural structure that will continue to be detailed, clarified, and sharpened. We may eventually be able to achieve a complete isomorphic item to item mapping.

The dilemma of consciousness will, nevertheless, still remain. Despite connection ...maybe isomorphic, full and complete, there will be no reduction that allows the external quantitative observation to enter into the internal qualitative subjective experience of consciousness. That is, not without the sharing of neural circuits by two or more individuals.

Humans do not have this capability, nor at the present, this technology. It is an empirical question, however, that we must leave open. Our species may someday achieve it. At the time of this writing, the only creatures thus far demonstrating it are Vulcans (i.e. Spock) of the science fiction television series Star Trek.

Chapter 23

FURTHER THOUGHTS ON PHYSICS AND THE PRIMARY ALGORITHM

The task of science is both to extend
the range of our experience and reduce it to order
(Niels Bohr, Nobel Prize, physics 1922) (Bohr 1934: 1)

Since the primary algorithm was initially defined in chapter 3 from the viewpoint of classical physics, it seems appropriate to present some further thoughts on the physics connection. The primary algorithm can be understood not only in terms of the laws of thermodynamics as discussed in chapter 3, but also in terms of the categories of time-space and energy-matter. Information is created, and information processing is made possible, by time-space, energy-matter differentials.

Simply put, the primary algorithm is at bottom a syntax...a generative, structuring algorithm, grounded in the genetic (somatic) informational matrix and including cybernetic evaluative and control mechanisms, expressing the combined alternative perspectives of the energy-matter and time-space unities. The unadorned graph of the hierarchical tree, which is the simplest graphical representation of the primary algorithm, illustrates the identity with these unities of physics.

In figure 23-1 that follows, the master node is being directionally disassembled into its parts, in the manner of deduction. Then, by mirror image the parts are being reassembled into the synthesis or master node inductively. The arrows represent the flowing unity of time/energy since both are required for an information processing or computational entity. The master node and the parts represent the matter/points-in-space unity, which is also required for an information processing or computational entity. Grammatically speaking, the arrows represent ***verbs*** (expressions of relationship, of time/energy). The nodes, as whole and parts, represent ***nouns*** (matter, objects, concepts, points-in-space).

To illustrate this to yourself or to another person, try the indicated simple experiment based on figure 23-1.

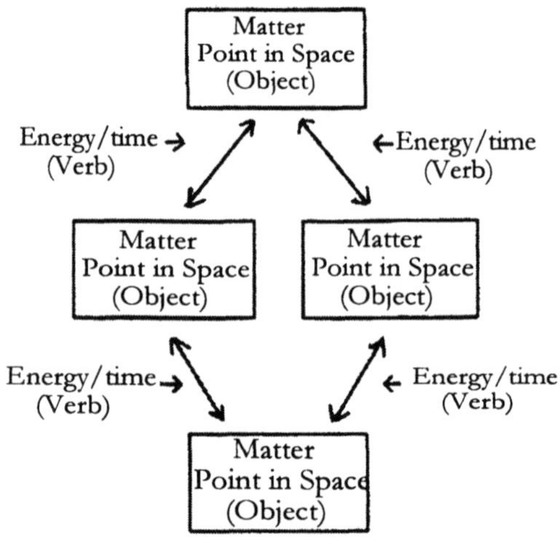

Figure 23-1. Hierarchical tree. Deductive/inductive, primary algorithmic mirror image (without showing the informational matrix containing cybernetic evaluative, control features, inhibitory and excitatory, necessary to the growth or generative algorithm)

Steps:

1. Give the syntactical algorithm any content that you like. That is, give your own choice of content to the root and the offspring, the parts and the whole.
2. Now explain the hierarchical tree and its relationships to a friend.

You will invariably use verbs to explain the arrow functions and nouns to designate the nodes (the root and offspring).

As the cognitive algorithm functions in our brains, it collapses space/time and energy/matter in the abstractions of our higher thought processes.

THE PRIMARY ALGORITHM AND QUANTUM PHYSICS

The expressions of the primary algorithm are nowhere more clearly seen than in the literature on the interpretation of quantum phenomena. A hybrid of the algorithmic forms ***order→ chaos→ order*** and ***incorporation→ extension→ incorporation*** is plainly seen in the introductory quote, repeated below, from the first words of the principal work on quantum physics by Danish physicist Niels Bohr, one of the pioneers of quantum theory.

> The task of science is both to ***extend*** (emphasis mine) the range of our experience and reduce it to ***order*** (i.e., incorporate) (emphasis mine) (Bohr 1987[1934]: 1).

The ostensible reality of quantum physics, as interpreted by major theorists, rests not on a finite objective reality but rather on the algorithmic processes of the brain. This is indicated by the following lines from Werner Heisenberg:

> The conception of the objective reality of the elementary particles has evaporated in a curious way...into the transparent clarity of a mathematics that represents no longer the behavior of the elementary particles but rather our knowledge of this behavior (Heisenberg 1958: 100).

In writing on the Copenhagen interpretation of quantum physics, physicist Henry Pierce Stapp, makes this point even clearer:

> But ideas and external realities are presumably very different kinds of things. Our ideas are intimately associated with certain complex, macroscopic, biological entities--our brains--and the structural forms that can inhere in our ideas would naturally be expected to depend on the structural forms of our brains. External realities, on the other hand, could be structured very different from human ideas. Hence there is no a priori reason to expect that the relationships that constitute or characterize the essence of external reality can be mapped in any simple or direct fashion into the world of human ideas. Yet if no such mapping exists then the whole idea of 'agreement' between ideas and reality and external realities becomes obscure (Stapp 1972: 1104).

Further on Stapp paraphrases and elaborates on the same thought expressed by Bohr. He writes:

> According to the pragmatic view, the proper goal of science is to ***augment*** and ***order*** (emphasis mine) our experience. A scientific theory should be judged on how well it serves to ***extend*** (emphasis mine) the range of our experience and ***reduce it to order*** (emphasis mine). It need not provide a mental or mathematical image of the world itself, for the structural form of the world may itself be such that it cannot be placed in simple correspondence with the types of structures that our mental processes can form (Stapp 1972: 1104).

The term ***augment*** carries the meaning of both extension and incorporation since it ***adds to*** the size or quantity. The terms ***extend*** and ***order***, as in the earlier statement by Bohr, are by this time readily identifiable with the primary algorithm. Stapp's statement, of course, tends to overlook the fact that the brain evolved as an interacting interface with its environment, and very likely, in the service of survival, maps that environment with reasonable accuracy. The brain is, after all, not a foreign product stuck randomly into this world from an alien universe, but rather is an adapted and functional part of this world (see the discussion in chapter 11 of this text).

The primary algorithmic process is illustrated by the contradictory ***wave function*** and ***real-particle*** approaches to quantum theory and their resolution. In the standard interpretation, the wave-particle duality of quanta is ***accommodated*** (in order to be ***incorporated*** into the quantum synthesis) rather than resolved. Simply put, the ***wave function approach*** moves up the hierarchy or level of generalization by defining the wave to be the ***inclusive*** mathematical representation (synthesis or generalization) that ***incorporates*** all the probabilities (parts, components, particles) of actual things that ***have*** occurred and therefore ***could*** occur as as outcome of measurements. The wave function is a high-level primary algorithmic synthesis that includes the probability of every possible event of the wave. The probability of some possible events may be very low, even remote...whereas the probability of others may be very high or expected. But all these events or things are included in the generalized wave function. Whatever does occur as an actual event, then, does so on the basis of its mathematical probabilities. The wave itself cannot be broken down further because it is engineered to contain ***all*** of the mathematical ***probabilities*** of ***things...not*** the ***actual things*** or ***events*** themselves. In contrast, the ***real-particle approach*** holds that there are not just probabilities but there are ***actual*** particles, very small localized objects, that stay together and do not spread out like waves.

The nature of the quantum dilemma, then, seems suggestive of the ***part*** and ***parcel*** aspects of the reductive/integrative alternatives of the primary algorithm. ***Wave*** is a integrative (holistic) concept; ***particle*** is a reductionist concept. Algorithmically, the same problem exists at all levels of thought... one can choose or prefer to view things from the standpoint of the whole or from the standpoint of the part, but it is very difficult to view satisfactorily from both perspectives at the same time. And the view ***is*** different from one algorithmic perspective or the other. The relationships between whole and part, as well as the empirical events that follow, may well seem different when viewed from the alternative perspectives. This is well expressed by Bohr, although he does not conceive of a primary algorithmic structure and process.

> ...the apparent contrast between the continuous onward flow of associative thinking and the preservation of the unity of the personality exhibits a suggestive analogy with the relation between ***the wave*** (emphasis mine) description of the motion of material particles ...and their indestructible ***individuality*** (emphasis mine) (Bohr 1987[1934]: 99-100).

As I noted in chapter 11, we tend to approach unity at both ends of the algorithm, the primal point and the end point. What has happened in both cases is, likely, that we have exhausted our powers of discrimination. Whenever we can discriminate no further, we tend to see unity or indivisibility, perhaps even a unity of probabilities. In other words the behavioral tension driving our bioneurological primary algorithm causes us great discomfort when we cannot unify or bring order to the chaos or parts. This discomfort of behavioral tension is probably the basis of psychological theories of cognitive dissonance and consonance.

In science, then, as well as in ordinary experience, to relieve the discomfort, we tend to impose unity or order upon the observations even to the point of accommodating contradictions such as wave-particle duality. The wave effect of chaotic phenomena seems to function similarly. Chaos theory deals with phenomena that are deterministic but not predictable. They are probabilistic, and are called nonlinear, in that the outcomes cannot be predicted from the known characteristics of their parts according to the so-called law of superposition (which means basically that the whole created is the sum of its parts). This may be merely an admission that we have not yet achieved the capability to discriminate accurately the parts and/or their relationships, or that the parts and relationships are too numerous for us to calculate and/or grasp. As a result, in order to make sense of our experience, we move primary algorithmically to an ***idealization*** (higher-level inclusive synthesis or generalization that is admittedly dubious by its very name; compare Stapp 1972: 1109; also compare the discussion of Bohr's concept of complementarity in Lowenstein 1999: 320, 331-332) that includes the sum of all probabilities or parts and ignores contradictions in the relationships among them that cannot be understood or resolved.

The famous dispute between Einstein and Bohr concerning determinism vs. indeterminacy in quantum physics may result in part from the fact that they proceed unwittingly from different perspectives of the algorithm. I say ***unwittingly***, because they have no concept of the primary algorithmic basis of their thinking, but rather a somewhat intuitive grasp of it. Einstein concentrated on the reductionist...the view from the part. Since the relationships between the parts were beyond our capacity to discriminate either instrumentally or algorithmically, he could not account for the seemingly contradictory parts of quantum phenomena. He therefore saw the quantum synthesis as lacking. The quantum theorists like Bohr and Heisenberg resolved their subjective discomfort by moving primary algorithmically to a higher level of synthesis, which established a new whole that included all the contradictory parts. Since they also could not discriminate and define the relationships, they fell back upon what, perhaps, may be considered an interim solution of assigning probabilities. Perhaps, at least in part, because of the inability to effectively discriminate, quantum theorists have formulated a holistic idealization which includes, but does not explain, the contradictions.

Shades of the yin and yang! And perhaps of Hegel. It may be justifiable to suggest the possibility that

To take quantum indeterminacy for reality may well be to mistake, instead, the shenanigans of our primary cognitive brain algorithm for physical reality.

Although this may appear a common and easy mistake in some writing, neither Bohr nor Heisenberg made that mistake. They realized that the quantum resolution rested not on the behavior in reality of the elementary particles but rather on our knowledge of this behavior.

Perhaps, the question of whether Einstein was right in his refusal to believe that God plays dice remains open when the nature of the primary algorithm is appreciated. Einstein also wanted the quantum result to reflect the objective reality of the physical world--i.e., he wanted the outcomes to be real. Bohr, on the other hand, was content to have the results not reflect actual reality but rather our ***knowledge*** of the resulting behavior. Einstein's approach was reductive and objective. Bohr's was pragmaticly integrative, but seemed to include also some aspects of subjectivity (e.g., see Penrose 1994: 309). Heisenberg, in contrast, at times inclined toward subjectivity.

The issue, however, may be resolved by the new advances in string theory. Recent string theory successes reportedly include the establishment of a inclusive framework that accounts for the previously illusive force of quantum gravity. String theory also offers an underlying organizing principle for the integration of general relativity and quantum mechanics, resolving the previous incompatibility between the two. String theory conceives of the fundamental units of nature not as point particles but as strings with miniscule degrees of length. The basic particles of quantum physics are proposed to be produced by the vibrational patterns of such ultramicroscopic strings.

The continuing, irrepressible unifying drive of the primary cognitive algorithm is well-expressed in the words of Harvard string theorist, Brian Greene, at the close of *The Elegant Universe* (1999):

> But even these paradigm-shaking discoveries are only ***part*** of a larger, ***all-encompassing*** (read *incorporating*) story. With solid faith that the laws of the large and the small should ***fit together*** into a ***coherent whole***, physicists are relentlessly hunting down the elusive ***unified*** theory (all emphases are mine) (Greene 1999: 386).

THE QUANTUM AMBIGUITY OF CONSCIOUSNESS

In an effort to grasp the seemingly quantum nature of consciousness and thought, we may go back to the mirror image of deduction-induction and the moving dialectic.

Conscious functioning suggests quantum duality or ambiguity at all levels of generalization/analysis. In fact, the feature of levels itself is always inherently dual or ambiguous. Each level contains all the parts or probabilities of its reduction or collapse. This may support the position of neuroscientist, Karl Pribram (1991), who sees consciousness as ***holonomic*** ...each ***part*** containing, or capable of generating, the ***image*** of the ***whole***. Further, in keeping with the flow of the primary algorithm, we tend always to assume or pursue another level of reduction or divisibility. Or the reverse, a higher level of inclusivity or integration. When we can no longer divide (discriminate, reduce, collapse) we confront this inherent duality most clearly. And we have the choice of imposing unity or else leaving the algorithm hanging in the state of chaos or uncertainty that was so uncomfortable to Einstein and others seeking a grand unified theory. These are also the choices reflected in the Kantian versus the Platonic dialectics.

RANDOMNESS AND PROBABILITY IN CONSCIOUSNESS

Randomness may or may not be inherent to consciousness, although it can certainly appear to be. We choose, at varying levels of conscious awareness, the actions, wave function collapses, reductions we make. Probability seems inherent, but not necessarily randomness. However, from the perspective of the outside observer, the reduction of conscious behavior may very well appear not only probable but also random.

Such is the nature of the primary algorithm. If the primary algorithm represents reality, it is because the algorithmic reality has been wired, by the process of evolution, into our cognitive functioning. On the other hand, if what we get is not the actual reality--then the representation we get must, in part or whole, be imposed by the algorithm.

The chaotic or extensive aspect of the algorithm facing up the scale of generalization is inductive...assembling the previously unrelated parts into wholes...into wave functions that contain all probabilities or parts. Facing down the scale of generalization, the chaotic or extensive aspect of the algorithm is deductive. The whole or wave function is reduced or collapses into it parts or its individual probabilities.

In thought...in the existing synthesis, generalization, or wave function of thought...all the parts, particles, or probabilities are present. But no individual part, particle, or probability has been selected, reduced, or collapsed. When one is selected...a course of action, a decision, or an observation has been made...the wave function reduces, collapses, etc., to only one of its probabilities or potentialities. Such is the nature of decision, action, or observation. Selection in thought can be viewed as deductive or reductive from the larger conscious wave function.

The possibility of finding a unified theory does not change the algorithmic nature of thought. We would, then, simply, reduce, divide, discriminate, or collapse further until we reach the point when the duality, or indivisibility (discrimination) is no longer possible. If we still perceive duality at that point, we will then tend to create unity of the duality. The creation of unity would then lead again to the chaotic algorithmic phase of questioning. It is noteworthy that string theorists already suspect the existence of parts more fundamental than their once fundamental strings (e.g., Greene 1999: 142). Greene seems to sense this when he speculates on the limits to explanation (1999: 384-386) and when he advises us earlier in his book:

> But history surely has taught us that every time our understanding of the universe deepens, we find yet smaller microscopic ingredients constituting a finer level of matter (Greene 1999: 142).

The algorithm seems inevitably to impel us to have unity, then chaos, then unity. Here we are back to the yin-yang or primal point--end point issue of the primary algorithm.

As noted in chapter 6, the spectrum of the conflict systems neurobehavioral model (with its tug and pull of ego and empathy) can also be thought of as analogous to a wave function including all the possibilities or probabilities of interpersonal behavior. It always collapses or reduces to one behavior in decision, action, observation. If it doesn't collapse, we see frustration, tension, indecisiveness, ambiguous behavior stalled in uncollapsed wave form. Externally, behavior is predictable from the model just like all quantum behavior…only on a probability basis.

The situation seems almost exactly comparable. But internally it is different. We know in our consciousness the tension, the difficulty, the struggle we go through in issues of greater ego and empathy. When we finally select an option, we collapse our conscious wave function into a specific act of behavior. In behavior, the behavioral wave function collapses into one part, particle, potentiality, observation, or other action. And that collapse becomes the actual, objectively observable reality of the behavior. ***Take, share, give*** are the three general reference points of the second algorithmic behavioral wave function, but the quanta of behavior actually spread over a range that we can only ***part***icularize approximately. I should also note that ***take, share***, and ***give*** are inevitably value choices based upon their somatic linkage. In our behavior, we make a ***choice*** from ***among values***, not detached from them.

The fact that the particular value choice, analogously to the second law of thermodynamics and quantum physics, cannot be predicted precisely but only on the basis of probability, leads to the erroneous conclusion of the separation of fact and value, i.e., the linguistic confusion that we cannot get from fact to value and vice-versa.

We may see a similar situation in mathematics. A math equation or formula can, likewise, be thought of as a constricted wave form or generalization in which the relationships of the parts are elaborated and expressed along the way to a full expression and solution as the equation is fully worked out.

When viewed from the perspective of a holistic, synthetic generalization or wave function, the independence of parts is illusory...as perhaps in a hologram. In the context of the whole, the parts cannot be independent...and a complete hologram can be generated from any part containing the algorithm of the interrelationships. They are inescapably interdependent.

The whole is broken, reduced, or collapsed when a part is considered as independent.

Viewed at a lower level of generalization or wave function, the focused upon part is then seen as independent, but ***its*** parts are not. They remain interdependent at that singular level and below.

THE BASIC QUESTIONS

At this point, we return to the basic, yet unanswered questions of chapter 11. These basic questions are: Has evolution provided us with a primary algorithm (of many variants) that shapes everything in its own image and prevents us from ever knowing reality as it truly is? Or have we been provided with a reality-matching or compatible algorithm that, employed rigorously, allows us to unlock the secrets of the universe? Or have we been given an algorithm that does a bit of both, giving us access to part of reality with the remainder forever beyond our ken? Despite our inability to clearly provide answers, the primary algorithm, as shown throughout the pages of this book, nevertheless, reveals itself to be ostensibly and, perhaps, substantially, isomorphic between behavior, thought, language, subjective experience, and, possibly even physical substrate.

Theoretical physicist and Nobelist, Murray Gell-Man lists four sources contributing to unpredictability and uncontrollability in quantum physics: 1. The fundamental indeterminacies of quantum mechanics; 2. The phenomenon of chaos; 3. The limitations of our senses; and, 4. Our indequate understanding and ability to calculate (Gell-Man 1994: 276). Perhaps we should add a fifth factor to the list...the interference, structuring, or limits imposed by our cognitive primary algorithmic processing capacity.

In conclusion, we may ponder a final question: How often are we re-vamping our theories, not to accommodate reality but rather our primary algorithmic capacity to relate our observations to its processing and structuring capacities and limitations? This is another way of asking how we may distinguish the order of reality from the order which our primary bioneurological algorithm imposes upon our perception and structuring of reality.

Chapter 24

CONCLUSION

...whenever empirical researchers discover enough new nonconforming phenomena to create chaos, synthesizers move in to restore order.
(Edward O. Wilson, Biologist, 1993: 243)

This book has been an effort at consilience, or the bridging of the natural and social sciences. The effort, anchored in physics, molecular biology, and evolutionary neuroscience, has sought to identify and trace the organic algorithms, which, replicated in our neural structure, shape our human thought, behavior, language, and subjective experience.

Figure 24-1. The primary algorithmic protein-nucleic acid synthesizing loop

I have identified two algorithms, or structuring, behavioral processes: the primary algorithm of organic synthesis (figure 24-1) and the second or reciprocal algorithm of evolved brain structure.

The primary algorithm was identified in the transition from nonliving matter to living organism. It is the fundamental algorithm of the protein-nucleic-acid synthesizing loop. The loop is a self-replicating, matrix-to-product-to-matrix circuit, in which the genetic informational matrix, uptaking and incorporating its information, materials, and energy, from the environment, builds a behaviorally active essentially protein structure, which, in turn, promotes the replication of the genetic matrix. It is an ordering algorithm with cybernetic features that reaches into and incorporates materials from the environment to maintain its ongoing vital processes. This algorithm is replicated in the cognitive and emotional structures and processes of our brain. The algorithm is expressed inductively and deductively in the familiar biochemical processes of anabolism (building up) and catabolism (breaking down). It is no accident, then, that the inductive/deductive framework of our cognitive processes is rooted in and displays the basic features of biochemistry.

I suggest that our cognitive inductive/deductive processes should be thought of as homologues of the basic organic structuring and information processing sequences. Increasingly we are finding homologies (structures and functions of similarity, correspondence, or identity based on continuity of genetic information) not only at higher levels of anatomy and behavior but also at the molecular level in the genome...from early platyhelminth-like worms of over 500 million years ago, through fruitflies, vertebrates, and up to human beings (e.g., see Smith, forthcoming; Galis 1999; Gehring 1998; Arthur 1997). Even the genetic code itself is a homology shared by all animal phyla (e. g., Arthur 1997: 28).

Succinctly stated, then, the essential inductive/deductive features expressed in our cognitive functioning were established at the time of the primal closure of the self-replicating, self-maintaining molecular protein-nucleic acid synthesizing loop and have been carried forward in the genome as basic features of our genetic informational matrix ever since. The process is appropriately acknowledged as the primary algorithm.

Some of the various terms expressive of this homologous or homeotic algorithm (with essential cybernetic features largely unexpressed or implicit), as manifested throughout the history of our thought and experience as well as across the range of our various academic disciplines, are as follows:

From physics

ORDER→ CHAOS→ORDER

From biology

ANABOLISM→CATABOLISM→ANABOLISM
INCORPORATION→EXTENSION→INCORPORATION

From cognitive psychology

EQUILIBRATION
EQUILIBRATION→ / →EQUILIBRATION
IMBALANCE

From Western philosophy

THESIS
SYNTHESIS→ / → SYNTHESIS
ANTITHESIS

From Mysticism West

CHANGE
PRIMAL UNITY→ (increase of multiplicity) →ABSOLUTE UNITY

From Mysticism East

CHANGE
COSMIC UNITY → (increase/decrease of multiplicity caused by tension of opposites) → COSMIC UNITY

From linguistic theory

SUBJECT
/
(THOUGHT) →(VERB) →SENTENCE
/
OBJECT

DISCRIMINATION (PART 1)
/
DISCRIMINATION (PART 1) →ESTABLISHMENT OF RELATION →DISCRIMINATION (SYNTHESIS 1&2)
/
DISCRIMINATION (PART 2)

From computer science and algorithmics as the hierarchical tree

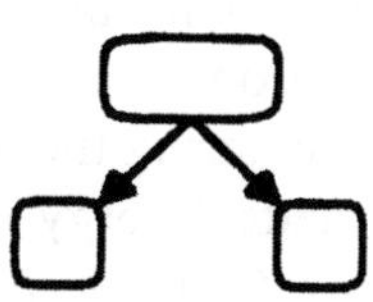

Figure 24-2. The typical upside-down binary tree with root at top and two offspring.

From logic, inductive and deductive, as a freeze frame or still photo:

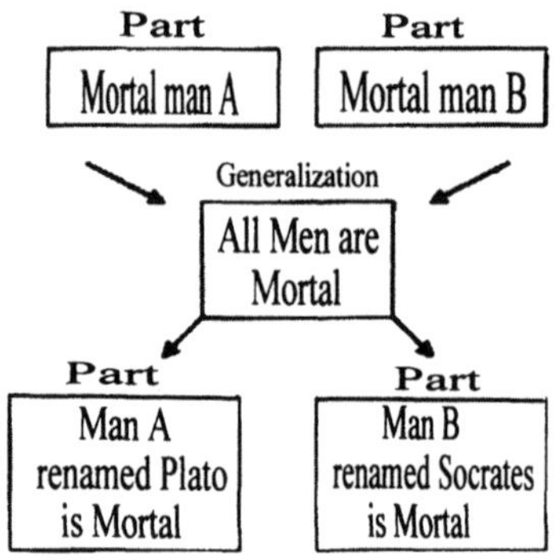

Figure 24-3. Deductive and inductive reasoning as mirror images.

And from the deductive/inductive freeze-frame turned sidewise and given movement:

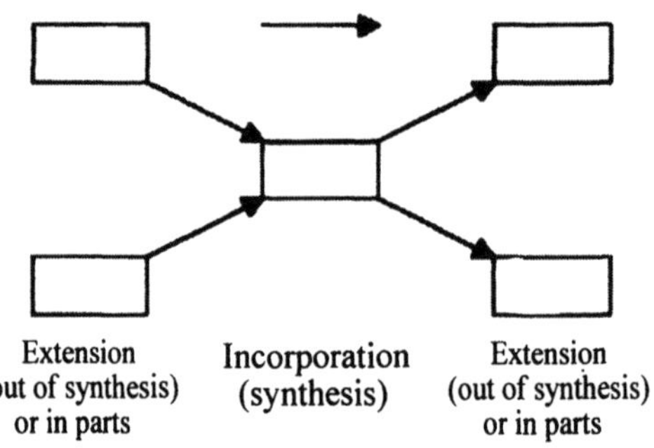

Figure 24-4. The mirror image deductive/inductive graph turned sidewise and given movement becomes the triadic form of the dialectic.

The organic synthesizing loop begins as the molecular self-promoting loop governed by the four basic forces of nature and the regularities of biochemical interaction. It follows the laws of thermodynamics...most conspicuously the second law. The molecular organic loop provides a unifying linkage between physics and biology. As further illustrated in chapter 23, the primary algorithm also expresses the space/time, energy/matter features emerging from classical physics, as well as general relativity, quantum mechanics, and string theory. In doing so it provides us not only with the space/time, matter/energy differentials necessary to any concept of information, but also with the derivative basic pattern, syntax, or grammar of thought, behavior, subjective experience, and language.

The second algorithm appeared much later. It began to evolve in the brain structure of mammals and emerged fully with the advent of that most advanced mammal, the human being, with its enormous brain capacity.

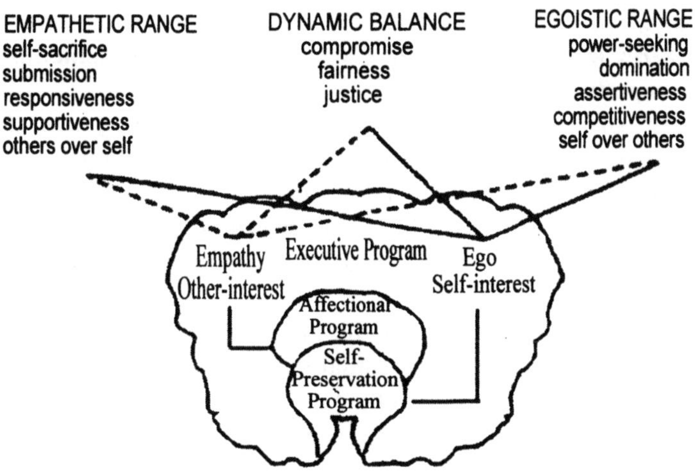

Figure 24-5. The major ranges/modes of incorporative/reciprocal behavior.

The two algorithms function together to form both the syntax and semantics of our behavior, our thought patterns, our meaning, our moral challenges, and their transformation into language. The primary algorithm is proposed as the fundamental structuring framework of our long-sought universal grammar. The highly generalized second algorithm of self-preservation and affection, ego and empathy, with all of its many composite subroutines functioning at all levels of our evolved behavioral processes, provides the fully articulated somatic matrix. As the behaviorally active expression of our genetic informational matrix...the evaluative or valuing component of all our experience, it assures the cognitive-somatic or mind/body linkage. All learning to include all lexical items of language (except, perhaps tag-along or incidental trivia) are coded to the various subroutines of this algorithm that operate through limbic and protoreptilian subcortical structures for emotional charge and significance. All subjective, as well as objective, meaning of language derives from this coding.

The two algorithms have been with us, shaping our lives since the dawn of history. Their articulation has been varied and slow. The two have been confounded with each other resulting in much intellectual confusion and in the unnecessary isolation from each other in the various fields of inquiry...academic disciplines. Some disciplines outwardly involve only the first algorithm in their methodology...although the second algorithm is present to influence methods in the person of the scientist or observer.

For instance: Pure logic uses the skeletal primary algorithmic framework and seemingly involves only minimally the second algorithm. That is probably a major reason why the sciences using pure logic...math, physics...have succeeded and seem so "hard", so clean, so relatively free of feeling and subjectivity.

It is an artificial detachment from the somatic cybernetic evaluation and control features which permits the Cartesian illusion of detachment from the somatic matrix.

The "softer" disciplines, the biosocial sciences, however, must deal more fully with the second algorithmic matrix. Humans and human systems are their subject matter. Because we have to account for the second algorithm as well as the first in the disciplines, of psychology, sociology, political science, and in the issues of philosophy, mysticism, and theology, they become very much more difficult.

This difficulty is compounded by the fact that the two algorithms are not distinguished but are always conceptually confused. Everything becomes even more messy because the results are evaluated by the standard of the logic of the first algorithm, which is seemingly easily externalized with minimal subjective, affective, and somatic components in the science of math and quantification.

Perhaps this analysis has contributed to the remedy of this situation so that we may more clearly confront the basic unity of all inquiry...to include the basic unity of fact and value in our behavior, experience, thought, and language. The identification, definition, and tracking of the two algorithms from physics through biology through neuroscience, provides a skeletal framework for demonstrating consilience…the unity of knowledge. Spelling out the linkages of this unity in full detail across the spectrum of the natural and social sciences may well occupy us throughout the next millenium.

CONSCIOUSNESS, THE PRIMARY ALGORITHM, AND THE BRAIN

It is hard to leave the topic of brain algorithms without a final speculative word about the increasingly popular subject of consciousness...the resolution of which will ultimately be necessary to a full achievement of consilience. More and more it is being considered, by serious researchers, that this most illusive but undeniable quality is not a discontinuous phenomenon, suddenly emergent and unique to humankind. But rather, consciousness, as we experience it, may well be a functionally continuous algorithmic elaboration rooted in the sensory-motor environmental responsiveness, and presumed sentient self-reference displayed by even the earliest single-celled bacteria (e.g., see Margulis and Sagan 1995: 32-33). British, molecular neurobiologist C. U. M. Smith, noting that the evolutionary tree reveals no discontinuities all the way down to the most basic organisms (prokaryocytes), wonders how far down the living world we can trace this inwardness, or sentient self-reference. He recoils emphatically, however, from the panpsychist view of consciousness at the level of atoms (Smith 1996: 471-472). Implicitly, then, Smith suggests that such inwardness must begin not lower than the level of the living organism; that is, at the level of self-replicating, self-maintaining protein-nucleic acid synthesis.

Our wide-ranging, multidisciplinary peregrinations indicate that all aspects of our thought, subjective experience, behavior, and language reveal the framework and process of the primary algorithm. Can we suggest, then, that the self-referential inwardness, leading ultimately to our self-aware consciousness, begins with the primary algorithm of biosynthesis? After all, the algorithm ***is***, conspicuously, the outward and visible origin of our subject/object relationship with the environment. Such a suggestion seems reasonable.[136]

Our living brain algorithms, then, proceed from this sentient, biological, and essentially protein behavioral substrate that maintains its unity cybernetically, as a functional organism, guided by its nucleic informational matrix, despite its spectacular and often very specific functional and structural ramifications. When the genetic informational matrix mutates or self-organizes, and then ramifies by confirmation of natural selection, it nevertheless retains what may be called the ***integrated modularity*** characterized by the organic synthesizing loop.

The brain, itself, functions as a complex and self-conscious unity, displaying such integrated modularity. It is comprised of pockets of modularity interwoven by complex, redundant, and recursive circuitry, allowing higher cognitive process to arise from the integration of such modularity. As noted, integrated modularity is based upon the primary algorithm of the organic synthesizing loop. The brain can be understood only in terms of both its parts and its unity.

The brain, therefore, likewise expresses the organic algorithm. Inductively, it assembles a unified consciousness from its parts. Deductively, it disassembles its unity into parts. If you surgically and

[136]The alternative viewpoint would be to view the primary algorithm as originating not in biosynthesis but later in cognition. As the structuring algorithm of cognition, its framework and process would then be ***projected*** or ***imposed*** on all aspects of our thought, subjective experience, behavior and language...even on what we observe, at the microbiological level, as biosynthesis. This alternative would virtually force us into the position of viewing all reality as an illusion structured by the cognitive algorithm. Knowing reality as it truly is, then, would be forever beyond our reach. Refer to the discussion in chapter 11.

artificially split its hemispheres from each other, some of its separate functional areas can be partially revealed. The working unity of the brain can be fully understood, however, only from the standpoint of overall ***algorithmic integrity***, which at the same time, simultaneously, grasps both whole and parts. Exclusive focus, or even overemphasis, on one algorithmic aspect, whole or parts over the other, inevitably is misleading or distorting. If you focus only on the unity, you obscure the obvious modularity. If you overly focus on the modularity, you likewise obscure the necessary unity.

In conclusion, I might again emphasize that before the appearance of the living, self-sustaining, self-replicating algorithm, there was no known comparable structuring process in nature. There is still no known alternative process. Despite enormous functional complexification and ramification, we still feel, behave, think, and speak in the framework and process, the syntax, of the primary algorithm of organic synthesis. To this primary syntax we add the subsequently evolved dynamic of our tri-level modular brain—the algorithmic somatic matrix driving the reciprocal tug and pull of our programming for self-preservation and affection, ego and empathy. Together, the algorithms function, not only to shape our behavior, thought, language, and subjective experience, but also to promote the survival of our individual selves and our species as a whole.

Appendix 1

MACLEAN'S TRIUNE BRAIN CONCEPT: IN PRAISE AND APPRAISAL

Paul D. MacLean is a pioneer, a trailblazer, a scientist, and thinker well ahead of his time.* As a humanist deeply interested in the larger questions of human life, he started out studying philosophy. Unable to find satisfactory answers to questions such as the origin and meaning of life...why humans in spite of their unrivaled intelligence, often behaved in seemingly irrational ways threatening their individual as well as species survival...he turned to medicine and the study of the human brain. He anticipated that the brain, as the biological substrate of these behaviors, held the key to better understanding of these fundamental questions as well as hopefully their answers. MacLean was, for many years, chief of the Laboratory of Brain Evolution and Behavior of the National Institute of Mental Health. In 1952, drawing upon the nineteenth century French scientist, Paul Broca's designation of the great limbic node which surrounded the brainstem of mammals, he introduced the conceptual term *limbic system* into the neuroscientific literature. In 1970 he introduced the concept of the triune brain, which became widely popularized after the publication of Carl Sagan's rather overly dramatic and simplified discussion of it in *The Dragons of Eden* (1977). MacLean, in further developing the triune brain concept, which aroused great interest in psychiatry, education, and the lay public, produced his detailed and highly documented volume, *The Triune Brain in Evolution: Role in Paleocerebral Functions* in 1990.

CRITICISMS OF MACLEAN'S MODEL

MacLean's triune brain concept has been acknowledged the single most influential idea in neuroscience since World War II (e.g., Durant in Harrington 1992: 268). Nevertheless, following the publication of his 1990 opus, MacLean received

* This appendix has appeared earlier in modified form in the ASCAP Newsletter, July 1998.

highly critical reviews in two prominent science periodicals, *Science* (October 12, 1990: 303-05) and *American Scientist* (September- October 1992: 497-98). Both reviews were written by neurobiologists and both reviewers claimed that MacLean's triune brain concept has had limited acceptance or been largely ignored by professional neurobiologists.[137]

Anton Reiner, at that time a recent graduate, wrote the review in *Science*, which was the more extensive of the two. After initially recognizing MacLean as a trailblazer of neuroscience, whose triune brain concept has been well-received outside the field of brain research, as the centerpiece of Sagan's popular, *The Dragons of Eden*, and frequently as the only discussion of brain evolution in psychiatry and psychology textbooks, Reiner makes several points in criticism of the triune brain concept.[138]

First, he notes that since MacLean introduced the concept, there has been tremendous growth in neuroscientific research that has greatly extended our knowledge of brain function and evolution. This statement, of course, carries the general implication, which Reiner later makes explicit, that the concept is out of date.

Secondly, in initiating a criticism of MacLean's concept of the limbic system, Reiner writes: "MacLean's presentation of the role of the hippocampus in limbic functions is not well reconciled with the current evidence that the hippocampus plays a role in memory."(1990: 304).

Thirdly, Reiner contends that current research indicates that MacLean's reptilian complex is not a reptilian invention but seems to be present in vertebrates all the way back to jawless fishes.

Fourthly, Reiner maintains that MacLean overreaches the evidence when he claims that the basal ganglia are the neural seat for the control of species -typical types of behaviors.

Fifthly, Reiner states that the limbic system, which widely used term MacLean authored as a pioneer neuroresearcher, is not properly represented by MacLean. Contrary to MacLean, as Reiner would have it, the limbic system did not appear first in early mammals. Amphibians, reptiles, and birds also have limbic features such as the septum, amygdala, a different-looking hippocampal complex, and maybe even a cingulate cortex.

Sixthly, Reiner asserts that MacLean assigns the functions of parental behavior, which Reiner claims that MacLean regards as uniquely mammalian, to the mammalian cingulate cortex, ignoring the fact that some reptiles (crocodiles), all

[137]For a highly favorable review of MacLean's 1990 book see the review by Emre Kokmen, M.D. of the Mayo Clinic, Rochester, Minnesota, in *J. Neurosurg*. V. 75, Dec, 1991, p. 998. In this appendix I focus on the reviews in *Science* and *American Scientist* because they have reached a wider audience and have become red flag reviews unjustifiably inhibiting the thoughtful application of the triune brain concept in related fields as well as in the psychological and social sciences.

[138]The criticisms made by Reiner are not necessarily in the exact order presented.

birds, and possibly even some extinct reptiles (dinosaurs) also engaged in parental behavior.

Seventhly, Reiner makes a couple of other criticisms of MacLean concerning a) his preference for correspondence over the more evolutionarily appropriate concept of homology and b)his apparently uncritical acceptance of Haeckel's idea that ontogeny recapitulates phylogeny.

Finally, although Reiner praises MacLean's motives and acknowledges the appeal of the triune brain concept for dealing with "big" behaviors that we are all interested in such as: "How does our animal heritage affect our behavior? Why do we do the things we do? Why can we not live together more harmoniously?"... he feels that there are some telling shortcomings as recited above, in MacLean's scholarship. He concludes that "neuroscience research ***can*** (emphasis mine) shed light" on these important human questions, "though ***perhaps*** (emphasis mine) not in as global and simple a way as MacLean has sought."

A CRITIQUE OF REINER'S CRITIQUE

Book reviews because of their very nature are usually overly brief. They usually cannot deal in depth with the points they take issue with. Reviewers, then, are often themselves guilty of the same kinds of oversimplifications and misinterpretations that they seek to expose in their reviews. When Reiner states..."I strongly believe the triune-brain idea to be wrong"... he is caught up in the same oversimplifying tendency that he claims unjustifiably to find troublesome in MacLean.

The triune-brain concept may be wrong in some of its particulars, right in others, but still be very useful and valid in its more general features. After all, at this stage of our knowledge of the brain although it is quite advanced over the 1960s and 1970s, there are not a great number of things we can say with absolute confidence...very few generalizations that are without arguable interpretations of more detailed research data. And Reiner takes apart but does not offer a replacement generalization. His analysis is destructive, not constructive. This type of analysis is the easy part of the job...almost anybody can do it.

But in his apparent eagerness to discredit and take apart MacLean's useful generalization, he also fails to study his subject closely and therefore engages in some very careless scholarship. He makes significant omissions, outright errors, and substantial misrepresentations of MacLean's work. Let's look at the points Reiner raises one by one.

1. ***Reiner blatantly misstates the facts when he claims that the triune brain concept as well as MacLean's book is outdated and lacks up-to date documentation.***

Reiner's first point i.e., that there has been a great growth in knowledge about the brain since MacLean first announced his triune brain concept in the 1960s and 1970s implies that MacLean has left the concept untouched and undocumented since that time and has therefore not considered any of the more recent findings. The implications of this statement are belied by the currency of research cited by MacLean and included in his discussions. In backing up his case for the alleged outdated ideas and data in the book, Reiner baldly states "only a handful of papers from the '80s are cited"(Reiner 1990: 305). This statement is categorically false and easily contradicted by a count of bibliographic items. The bibliography of this work contains over 180 entries (a big handful indeed!) which date from 1980 to at least 1988 and over 220 entries that date between 1975 and 1979. This amounts to at least 400 entries of rather recent documentation...keeping in mind, of course, that the publication date of MacLean's book and Reiner's review was for both 1990.

2. ***Reiner ignores or misstates the facts when he says, "MacLean's presentation of the role of the hippocampus in limbic functions is not well reconciled with the current evidence that the hippocampus plays a role in memory."***

The phrasing of this statement indicates that MacLean is unaware of or fails to report on the extensive research indicating the role of the hippocampus in memory. Such an implication is totally unwarranted. MacLean devotes fully two chapters to reporting and discussing such research. These chapters even have memory in their titles. Chapter 26 is titled *Microelectric Study of Limbic Inputs Relevant to Ontology and* ***Memory*** (emphasis mine). Chapter 27 is titled *Question of Limbic Mechanisms Linking a Sense of Individuality to* ***Memory*** (emphasis mine) *of Ongoing Experience.* These chapters deal at length with the role of the hippocampus in memory and propose an integrative role for the hippocampus in tying learning to affect or emotion (For a summary of MacLean's discussion on these matters, consult 1990: 514-16).

3. ***Claiming that the reptilian complex is not a reptilian invention, Reiner misrepresents MacLean's position.***

On the third point, Reiner contends that current research indicates that MacLean's reptilian complex is not a reptilian invention but seems to be present in vertebrates all the way back to jawless fishes. This is largely a taxonomic question. At what point do we declare something to be a fish, an amphibian, an amniote, a reptile, or a mammal? And do we view mammals as branching off from the amniote tree before we have distinct reptiles in the line of descent? Or do we prefer the more likely probability that mammals descended in a line from the ancient mammal-like reptiles of the pre-dinosaur Permian-Triassic periods called therapsids, who represent a branching of the ancient reptile line (cotylosaurs). Therapsids appeared approximately 230 millions years ago, and approximately 50 million years before the emergence of the great dinosaurs of the Jurassic and Cretaceous periods.

MacLean knows these facts and clearly acknowledges them, while supporting a lineage for mammals that traces back to the therapsids, of the synapsida subclass that branched off from the diapsida line that eventually produced the great dinosaurs many years later. This is the standard position in evolutionary theory today. One might wish to compare the phylogenetic tree in MacLean (1990: 34) with Butler and Hodos (1996: 72), Strickberger (1996: 396) and Hickman, et al. (1984: fig. 27-1). And it is the accepted position of standard zoology texts (e.g., Miller and Harley 1992, Hickman, et al. 1984, 1990). Mammals, and ultimately us humans, then, did not evolve from dinosaurs but from a parallel lineage that split much further back in geologic time.

If the term Reptilian Brain or Reptilian Complex causes confusion with modern reptiles, and because the reviewers don't wish to read MacLean's work closely, the Reptilian Complex could be thought of, and perhaps redesignated, as the ancient amniote complex or even the early vertebrate complex. And, of course, as MacLean acknowledges thoroughly, this early brain complex is not the reptilian brain of modern reptiles but it is also not the same as that of the early vertebrates, amniotes, or therapsids. At several points in his book, MacLean makes this unequivocally clear by his reference to stem reptiles (cotylosaurs) (MacLean 1990: 33, 82), those early reptiles from which both the diapsid and synapsid lines branched off. To assure the proper evolutionary context, MacLean also uses the term *protoreptilian* in his initial definition and adds the clarifying comment that he refers to the reptilian complex (or R-complex) only for brevity's sake (see MacLean 1990: 15-16, 244, 519). This protoreptilian, or stem reptile brain, has been altered by modifications which include those produced by differentiation and elaboration of earlier structures (e.g., see MacLean 1990, 243). These modifications, to include differentiations and elaborations, provide, in addition to their previous maintenance and behavioral functions, neural circuitry in support of the enhanced limbic structures of mammals. These enhanced mammalian limbic structures necessarily engage and enhance prior circuitry in the brainstem. And together these enhanced limbic and brainstem circuits provide support for the greatly enhanced neocortex (or isocortex) which eventually got the sufficient modifications that permitted language and the development of complex technological societies.[139]

4. ***Reiner misrepresents MacLean's position on the basal ganglia.***

On the fourth point above, Reiner states that he knows of no one other than MacLean who believes the basal ganglia to be the neural seat for the control of species-typical types of behaviors (Reiner 1990: 305). This statement is a misrepresentation of MacLean's position as well as an admission of ignorance on the part of Reiner. In the first place, MacLean never uses the inclusive term "neural seat." Further MacLean is not talking about all species typical behavior but only some. He specifically excludes from this discussion such mammalian class/species typical behavior as maternal nursing and play, which are attributed primarily to other brain parts and treated in other chapters of the book.

In part II on the *Striatal Complex with Respect to Species-Typical Behavior*, MacLean repeatedly emphasizes that the traditional view that the striatal complex is primarily involved in motor functions represents an oversimplification. He writes that the purpose of the present investigation is to test the hypothesis that the striatal complex plays an "essential" role in certain species typical behaviors as well as certain basic forms of behavior common to both reptiles and mammals (MacLean 1990: 243). At one point after reciting the evidence, MacLean says that the results "suggest that the medial globus pallidus (a structure of the basal ganglia) is a site of convergence of neural systems involved in the species-typical mirror display of gothic-type squirrel monkeys." (MacLean 1990: 189). And, a little further on, that "findings indicate that in animals as diverse as lizards and monkeys, the R-complex

[139]The use of the term "additions" is deliberately avoided here because it has been the source of some confusion (see Butler and Hodos 1996: 86). New brain structures do not spring de novo out of nowhere but rather evolve from the differentiation of previously existing structures. When differentiations become sufficiently established, they are often referred to loosely as "additions." This does not deny that seemingly new additions may possibly and occasionally arise, but the intent here is to emphasize the phylogenetic continuity that underpins the concept of homology

is ***basically involved*** (emphasis mine) in the organized expression of species-typical, prosematic communication of a ritualistic nature." (1990: 189).

Additional research, some predating some postdating Reiner's review and of which Reiner is apparently ignorant, adds further support to MacLean's hypothesis. For example, J. Wayne Aldridge and colleagues from the University of Michigan in a research report titled "Neuronal Coding of Serial Order: Syntax of Grooming in the Neostriatum,"(1993) conclude that there is "direct evidence that the neostriatum ***coordinates the control*** (emphasis mine) of rule-governed behavioral sequences." This study builds upon a series of earlier studies of species-typical grooming behavior of the rat (e.g., Berridge and Fentress 1988; Berridge and Whishaw 1992; Cromwell and Berridge 1990). These earlier and more recent studies certainly support MacLean's hypothesis that the striatal complex plays an essential role in some species typical behaviors of a ritualistic nature.

And, of course, there is the growing body of clinical evidence, going well back into the 1970s and 1980s, that neurological disorders in humans (such as Parkinson's, Huntington's, and Tourette syndromes), which involve damage to the neostriatum, produce specific deficits in the sequential order of movement, language, and cognitive function.(e.g., Holthoff-Detto, et al. 1997; Cummings 1993; Benecke, et al. 1987; Marsden 1982, 1984; Oberg and Divac 1979). Such serial order patterns in behavior are phylogenetically old as well as pervasive and often constitute the basis of identifying so-called species-typical behaviors.

5. ***Reiner misrepresents the facts when he claims that MacLean says the limbic system first appeared in mammals.***

On the fifth point above, Reiner again misrepresents MacLean's position. MacLean does not claim that the limbic system first appeared in early mammals. He acknowledges that limbic features appear in fishes, reptiles, and birds, but are rudimentary and poorly developed as compared with those of mammals (MacLean 1990: 247, 287). According to MacLean's view, then, it is not the presence or absence of limbic features themselves in ancestral amniote or reptilian vertebrates, but rather the significant and prominent development of limbic features in mammals which is appropriately of interest in understanding the evolution of characteristically and uniquely mammalian behavior. Further, care must be exercised in making comparisons across existing modern species. We can only infer that the structures and undeveloped and/or rudimentary homologues of such structures in modern species were also present in ancestral lines. Brains don't fossilize, so the point can't be made conclusively. The currently accepted inferential position in neuroscience is that there are homologues of limbic structures going well back into vertebrate history, although these homologues in modern species are often difficult to establish and sometimes downright dubious (Striedter 1997; Veenman, et al. 1997).[140]

[140]The accuracy and utility of the concept and term limbic system has itself been a separate topic of some disagreement in recent years. Some authors state that it does not represent a truly functional system and the term should be discarded. Others defend its use. Most texts continue to find the term useful and because of its longtime usage it will probably remain in the literature. Some recent and prominent scholars illustrate the controversy well. Pierre Gloor of the Montreal Neurological Institute, McGill University, in his thorough-going work *The Temporal Lobe and Limbic System*, by the very use of the term in the title indicates his position. Further on in the text, while acknowledging the controversy he writes that this

6. ***Reiner displays careless scholarship and misrepresents the facts of neuroscience, evolution, and animal behavior as well as MacLean's position on parental behavior and the cingulate cortex.***

Another example of careless scholarship on Reiner's part is in the sixth point above. He claims that MacLean assigns the functions of parental behavior, to the cingulate cortex and that MacLean regards parental behavior as uniquely mammalian. According to Reiner, MacLean's alleged position "ignores the fact that some reptiles, such as crocodiles, and all birds engage in parental behavior, not to mention the possibility suggested by paleontological data that some extinct reptiles, namely dinosaurs, also engaged in parental behavior."(Reiner 1990: 305).

Such a blanket claim suffices to make one wonder if Reiner felt it worth his while to even consult the book he is reporting on. Firstly, MacLean does not "assign" parental behavior to the cingulate cortex. He reports the recent (at that time) research on maternal mechanisms in the septal or medial preoptic area (MacLean 1990: 351-53) and indicates that this area may have provided the initial potentiality for full scale mammalian maternal behavior (MacLean 1990: 354), which would include play and the development of empathy. The very title of his chapter 21 is ***Participation*** (emphasis mine) *of Thalamocingulate Division in Family-Related Behavior.* Participation is participation not unilateral and unequivocal assignment. And MacLean uses the systemic term thalamocingulate to indicate intra-limbic nuclei and cortical connections, not simply cingulate cortex as Reiner states.

---

system in mammals exhibits an organization that is sufficiently different from that characterizing other areas of the cerebral hemisphere to merit such a designation (Gloor 1997: 106).

And well-known neurologist, Richard Restak tells us that based upon a large body of experimental work, it is appropriate to conclude that, "depending on the areas stimulated, the limbic system serves as a generator of agreeable-pleasurable or disagreeable-aversive affects."(1994: 143). Nevertheless, there is little agreement among neuroscientists concerning the contributions of the different components, and their mutual influence on each other (1994: 149).

On the other hand William Blessing, a neuroscientist at Flinders University, in his study of the lower brainstem, feels that emphasis on the limbic system has detracted from the study of brainstem mechanism, that it has been "plagued by its anatomical and physiological vagueness and by the lack of precision with which the term is used."(Blessing 1997: 15). Further, he feels the term should be dropped from the literature(Blessing 1997: 16).

A third recent author, neuroscientist Joseph LeDoux (1996: ch. 4) argues that because the limbic system is not solely dedicated to the single global function of emotion, a claim that MacLean fully recognizes in his chapters on memory (1990: chs: 26 & 27), that the concept should be abandoned. LeDoux apparently prefers a single functional criterion for the definition of a system, whereas MacLean seems to prefer a combination of functional and anatomical criteria. Le Doux concludes his argument by stating: "As a result, there may not be one emotional system in the brain but many."(1996: 103). Compare this with the concluding line of the definitional description by Kandel, et al., authors of the most widely used textbook on neuroscience and behavior: "The limbic system contains neurons that form complex circuits that play an important role in learning, memory, and emotion."(1995: 708).

The use and value of the conceptual term limbic system, then, seems to depend on one's research focus and how one chooses to define a system. It might be added that the definition of what constitutes a system is controversial in all disciplines, not just in neuroscience.

MacLean cites good evidence for thalamocingulate participation in "nursing, conjoined with maternal care"(MacLean 1990: 380). After all, lesions in certain portions of the cingulate cortex interfere with nursing and other maternal behavior (Stamm 1955, Slotnick 1967), not with blanket parental care as Reiner asserts.

It may be too early or simply erroneous in neuroscience to assign anything specifically and finally to any exclusive part of the limbic area. There is more likely some localization of minor function, but for most behaviors of any scale there seems to be fairly wide-ranging neural circuitry that may be interrupted by lesions at many different points. For example, recent research on maternal behavior (nursing, retrieval, nest-building) in rats has focused on the medial preoptic area with its connections to other limbic structures and the brain stem (Numan 1990). Alison Fleming and her colleagues (1996), summarize what we know about the neural control of maternal behavior. Not only the medial preoptic area with its brainstem projections, but also other limbic sites are involved, including the amygdala (Numan, et al. 1993; Fleming, et al. 1980), hippocampus (Terlecki and Sainsbury, 1978; Kimble, et al. 1967), septum (Fleischer and Slotnik 1978), and cingulate cortex (Slotnik 1967, Stamm 1955). Most emotions, emotional behaviors, and emotional memories seem to be distributed, involving multiple pathways. Specific behaviors and categories of behaviors can be interrupted by lesions at varying points in these multiple pathways. More recent research has again confirmed that the cingulate cortex is involved in emotion and motivation (Stern and Passingham 1996). In a recent research report John Freeman and colleagues conclude that the neural circuitry formed by interconnected cingulate cortical, limbic thalamic and hippocampal neurons has fundamentally similar functions in the affective behaviors of approach and avoidance (Freeman, et al. 1996).

Like any good scientist with an open mind, MacLean, at the close of his chapter on participation of the thalamocingulate division in family-related behavior, calls for more neurobehavioral research to explore the extent of this participation (MacLean 1990: 410). It is also noteworthy that MacLean is one of the few thinkers in neuroscience who shows concern for the neural substrate of such family based behavior, characteristic of mammals, as play and the underpinning but illusive quality of empathy. Although such characteristics have been reported on behaviorally (e.g., for play, see Burghardt 1988, 1984; Fagen 1981), they have largely been ignored in the search for neural substrates, not because they are unimportant, but because of the extreme difficulty in defining and objectifying them. But the evidence clearly points to neocortical as well as limbic cortical and subcortical representation (e.g., see Fuster 1997: esp. 169; Frith 1997: 98; Frith 1989: 154-55). One of these days, hopefully, mainstream neuroscience will direct more serious research toward a better understanding of these difficult and ignored questions which are so critical to a full understanding and appreciation of humanity.

Reiner also indiscriminately uses the blanket term "parental behavior" coupled with attributing that same blanket usage to MacLean. In this usage, Reiner shows a remarkable deficit of scholarship, naivete, or both. MacLean is not discussing all parental behavior. He is discussing those nurturing behaviors that are the most distinguishing characteristic of mammals and a fundamental part of their taxonomic classification and differentiation from birds and reptiles. These behaviors must be found in either new structures or modifications to existing structures. As Butler and

Hodos point out, new structures may be added to organ systems, but modification of existing structures appears to be more common (1996: 86). The jury is still out on the neurophysiology of these defining mammalian behavioral features. What's more, with the emphasis on cognition in neuroscience, there has been surprisingly little attention paid to the extensive work on the neural and hormonal basis of the motivational and emotional aspects of maternal care. This is openly acknowledged by leading scholars in the brain science field (e.g., Rosenblatt and Snowden 1996; LeDoux 1997: 68; Kandel, Schwartz, and Jessell 1995).

The blanket term "parental care" as used by Reiner in his criticism of MacLean amounts to condemnation by indiscriminate generalization. Parental care has been defined by a leading authority as "any kind of parental behavior that appears likely to increase the fitness of the parent's offspring" (Clutton-Brock 1991: 8). It is a very broad and inclusive term. The term includes nest and burrow preparation. The very production of eggs itself is included. This kind of "parental care" is found in the earliest vertebrates with very primitive brains indeed. If the all-inclusive definition of parental care can be stretched to include the production of eggs and digging a hole to place them in, perhaps it could conceivably be stretched to include even the sharing of cellular membranes during asexual reproduction by single-celled organisms.

But specifically...what about parental care in modern reptiles? Contrary to Reiner's claim, MacLean reports on parental care in crocodiles (MacLean 1990: 136-37) and also in some species of skink lizards (MacLean 1990: 136, 248-249). A recent review article on parental care among reptiles by Carl Gans of the Department of Biology, University of Michigan, brings us up to date. Gans claims that the most spectacular example of reptilian parental care takes place among crocodiles. Both parents respond to the call of hatchlings who vocalize underground while emerging from the eggs. The adults dig them up and transport them to water in their large buccal pouch(Pooley 1977). The young are then washed and stay shortly in association with the adults. After a relatively brief period, however, the juveniles' response to the adults reverses. The juveniles disperse suddenly into small, nearby channels where they may dig themselves tunnels. Gans notes:

> In view of the fact that crocodylians may be ***cannibalistic*** (emphasis mine), there seems to be both an inhibition of cannibalism in the parents and an inhibition of a possible adult avoidance reaction in the neonates (1996: 153).

This kind of short-lived parental care during which the cannibalism of parents is inhibited may be impressive in reptiles, but it is a far, far cry from the highly developed family-related behavior in mammals; behavior which is so further developed in the human species that it extends often throughout an entire lifetime and becomes the basis for a vastly extended social life. The ***equating*** of parental care in reptiles with parental care in mammals is simply ludicrous. It is this mammalian family behavior that concerns MacLean, and the neural substrate is appropriately sought in the brain modifications that became prominent with the appearance of mammals.

7. ***Reiner's further inaccuracies: recapitulation, homology, and correspondence, etc.*** Near the end of his review Reiner makes the following isolated statement: "MacLean also errs in his apparent sweeping acceptance of Haeckel's idea that ontogeny recapitulates phylogeny." Again Reiner distorts and

misrepresents. From a close review of the book it is by no means altogether clear that MacLean "sweepingly" accepts Haeckel's concept. In fact he only refers to it once (MacLean 1990: 46) while at the same time noting the well known exceptions. Haeckel's concept has been replaced in neuroscience today by the principles of von Baerian recapitulation. The von Baerian version holds that while ontogeny does not recapitulate phylogeny in the thoroughgoing Haeckelian sense, it does recapitulate the features of an organism in terms of the organism's general to more specific classification. In other words the von Baerian principles state that the more general features of an organism develop before the more specific features (Butler and Hodos 1996: 51-2). The issue, however, is still not so clearly settled. The emergent discipline of evolutionary developmental biology is looking more closely into such questions (Hall 1992, Thompson 1988). For instance, evolutionary biologist Wallace Arthur, in summarizing the main themes of this emerging discipline, writes:

> No single comparative embryological pattern is universally found or can be described as a 'law'. Von Baerian divergence, its antithesis (convergence) and a broadly Haeckelian (quasi-recapitulatory) pattern can all be found, depending on the comparison made (1997: 292).

On the additional point that MacLean prefers to think in terms of correspondence rather than homology probably reflects his functional-behavioral orientation. In fact it is specifically in discussing the issue of the relationship between structure and behavior that (MacLean 1990: 37) makes this comment. Later, he returns to a more standard use of homology (MacLean 1990: 228). There is, in fact, presently no sure fire way of demonstrating that homologues have the same one-to-one functions or produce the same one-to-one behaviors across species. In reporting that MacLean, at one point, expresses preference for the term correspondence because of the confusion in the definition of homology, Reiner shows what can only be considered a misplaced and sophomoric "gotcha" exuberance. He writes that MacLean's comment "should leave Stephen J. Gould, not to mention all other students of evolution, aghast," adding that such a comment constitutes a "very critical misjudgment to make in a work on evolution."(Reiner 1990: 305).

This is truly a naive, if not preposterous statement by Reiner. Could it be that Reiner is not aware of the long history of the pervasive problems associated with the definition of homology? For example, Leigh Van Valen, of the biology department of the University of Chicago, in the first sentence of his frequently referenced article on homology and its causes, writes: "Homology is the central concept of anatomy, yet it is an elusive concept."(1982: 305). Further on, in view of the persistent definitional ambiguities, Van Valen practically equates the two terms homologue and correspondence when he writes: "In fact, homology can be defined, in a quite general way, as ***correspondence*** (emphasis mine) caused by a continuity of information" ...although in a footnote Van Valen admits that correspondence itself needs further definition beyond the scope of his paper (305: fn. 1; cf. Roth 1994). Although there has been some sharpening of the concept of homology, with emphasis on phyletic continuity, the ambiguities have by no means been adequately resolved (Arthur 1997: 171-77; Hall 1994, 1996).

And there is the haunting question that is still wide open for research and investigation ...do most homologous behaviors share a homologous structural basis or can homologous behaviors be rooted in nonhomologous structures? (see Hall

1996: 29 fn. 23). The recent report by William Blessing on the lower brainstem raises the question of multiple neural representations of body parts and behavior, in that behavior originally represented and controlled in the brainstem of an earlier vertebrate may maintain its brainstem representation, but be controlled by an added representation in the frontal cortex of a more highly developed mammal. Such multiple representations at different levels as the brain became more complex would certainly confuse the issue of a straightforward homologous match of structure and function (1997: 1-18; see also, Brown 1977).

Research on very limited aspects of function are often suggestive but far from conclusive even on such limited function. Establishing homologues of the prefrontal cortex can be particularly vexing. A recent research article by Gagliardo and colleagues, "Behavioural effects of ablations of the ***presumed*** (emphasis mine) 'prefrontal cortex' or the corticoid in pigeons" (Gagliardo, et. al., 1996), indicates, not only in its discussion and conclusions, but in the very title itself, the uncertainty, ambiguity, and cautions that currently characterize such research efforts (see also Fuster 1997: 7-11).

There is an awful lot of assuming going on in some quarters of neuroscience on this issue, which simply cannot be settled at this time based on the empirical evidence. This is one of the problems and cautions that must be acknowledged when generalizing across species...say from rats to humans. In maternal behavior, for example, can we say factually that the medial preoptic area plays the same part in the maternal behavior of humans that it does it the rat brain? No, we cannot. At least not yet. But neuroscientists, after first hedging themselves, and following homologous logic, seem inclined to think so. Nevertheless, it is entirely within the realm of possibility that we may find that it does so only in part or not at all. As neuroresearcher Joseph LeDoux notes: "Some ***innate*** (emphasis mine) behavioral patterns are known to involve hierarchically organized response components." (1996: 120). And further on he adds: "Species differences can involve any brain region or pathway, due to particular brain specializations required for certain species-specific adaptations or to random changes."(1996: 123). And neurologist Richard Restak points out that in the case of animals multiple limbic areas may increase, modify or inhibit aggression. He notes further that even the same area may increase or inhibit responses under different experimental conditions and depending on the animal selected for experiment. As an example, he points out that the destruction of the cingulate gyrus (a limbic component) increases aggressive behavior in cats and dogs, whereas, on the contrary, such an operation has a calming effect in monkeys and humans (1994: 149).

Or perhaps, as Blessing suggests, there are multiple representations. Then we might have to go to correspondence rather than homology (even homoplasy might not apply, since homoplasy, or parallel evolution, would probably not apply in such closely related species) to account for the behavioral circuitry. In other words the corresponding neural circuitry--that circuitry controlling maternal behavior --may be found in the same, slightly differing, multiple, or perhaps (though highly unlikely) even totally different structural homologues or modifications.

In fact, if homology is correct and functionally, to include behaviorally, uniform...that is, the same structures account for the same functions and behaviors across classes, orders, and species... this finding would support the triune brain concept as set out by MacLean, which says generally that the protoreptilian complex common to both reptiles and mammals functions largely the same in both classes. This finding would also support MacLean's position that the expanded circuitry areas of the mammalian complex bear characteristically mammalian functions and are the circuitry for characteristically mammalian behaviors such as nursing, a defining taxonomic feature of mammals (which, in part distinguishes them from reptiles and birds).

In a final series of somewhat negatively gratuitous comments Reiner writes about some of MacLean's legitimate speculations. For example, Reiner states "...and mathematical skill (he thinks the cerebellum could be involved)"(Reiner 1990: 305).

And why not? See MacLean's discussion on the subject (MacLean 1990: 548-52). Recent research has indicated that the cerebellum is not just a motor mechanism, but is also likely involved in higher cognitive and perhaps even language function. Especially relevant is the rather well-supported hypothesis that indicates a cerebellar mechanism involved in all tasks that require precise temporal computations. This could well suggest an involvement in mathematical processes. True, the evidence is insufficient to permit firm conclusions as to the cerebellar role in higher cognitive processes, but it is a research direction which needs further refinement and is currently pursued by a number of neurobiologists (Daum and Ackermann 1995; Dimitrov, et al. 1996; Altman and Bayer 1997: esp. 749-51).

Overall, given the outright errors, careless scholarship, misrepresentations, and sophomoric, prejudicial tone of Reiner's review, it probably should never have been allowed to appear in a publication of the stature and influence of *Science.* Such prejudicial reviewing should perhaps raise serious questions of standards if not ethics in the academic-scientific community.

CAMPBELL'S REVIEW IN AMERICAN SCIENTIST

The review by Campbell in *American Scientist* (1992) is a much shorter review than that of Reiner. It brings up some of the same points, but is less prejudicial in its tone. Since it is less detailed it expresses primarily the preferences and value judgements of the reviewer. Campbell repeats Reiner's erroneous charge about outdatedness. He writes: "...that except for a very few papers, most of the references were published prior to 1980"(1992: 498). It has already been noted that this "handful" of items amounts to more than 180 citations. One suspects that Campbell proceeded from his preconceptions and found what he expected to find. Campbell ends his review with the statement: "Unfortunately, the data presented are, ***to some degree*** (emphasis mine), outdated, and the evolutionary reasoning is unsophisticated."(1992: 498). The use of the term "unsophisticated" by the reviewer is a good example of gratuitous abuse of review. It is a sweeping value-laden term that communicates more about the reviewer than the reviewed. For anyone who has closely read MacLean's detailed and thoughtful work, such blanket judgments are not warranted. The evolutionary reasoning is, on the contrary, quite thoughtful, well-presented, and sophisticated. Such blanket judgments tell us more about the sociology of neuroscience and neuroscientists that they do about the subject matter of the discipline itself.

THE COMMENTS OF BUTLER AND HODOS

In their recent comprehensive and overall admirable work on comparative vertebrate anatomy, Butler and Hodos attempt to formalize the assignment of MacLean's work to the relics of history. Their comments reflect the standard oversimplified criticisms, misrepresentations, and errors that have become popular to repeat ever more unreflectively. Butler and Hodos assign the triune brain concept, inaccurately and indiscriminately, to a category they called "theories of addition." And without any detailed discussion or analysis, of the very significant indisputable points of accuracy in MacLean's concept, they write that the past three decades of work in comparative neurobiology "unequivocally" contradicts MacLean's theory (1996: 86).

It seems almost incredible that two such qualified authors should accept the same flagrant misrepresentations, inaccuracies, and oversimplifications of MacLean's work that have become commonplace in some sectors of neurobiology over the past decade. It appears that they merely parroted the errors and misrepresentations of Reiner and others rather than reading MacLean's 1990 work closely and open-mindedly. Or perhaps they simply took their understanding from Carl Sagan's overpopularized and oversimplified presentation in *The Dragons of Eden* and didn't consider the issue worth looking into further. There is no point in repeating the responses given earlier to Reiner's review. The same points hold for Butler and Hodos' comments. The rebuttal points are clearly made and easily accessible to verification by anyone who chooses to make the effort. The categorical statement by Butler and Hodos that the extensive body of work in comparative neurobiology over the past three decades unequivocably contradicts MacLean's theory, which they apparently have not read, constitutes on that point poor, if not irresponsible, scholarship.

THE UTILITY AND VALIDITY OF MACLEAN'S TRIUNE BRAIN CONCEPT

The triune brain concept may have its faults. But such faults have been patently misrepresented in some cases and grossly exaggerated in others. Whatever its faults may prove to be, the triune brain concept gets at a fundamental truth. The mammalian modifications, differentiations, and elaborations to the early vertebrate and ancestral amniote brains had the effect of introducing endothermy (warm-bloodedness), maternal nursing, enhanced mechanisms of skin contact and comfort, as well as enhanced visual, vocal, and other cues to bond parents to offspring and serve as the underpinning for the extended and complex family life of humankind. The mammalian modifications, therefore, added greatly enhanced affectional, other-interested behavior to the primarily (although not exclusively) self-preservational, self-interested behaviors of ancestral amniotes and early vertebrates (not necessarily their modern representatives).

The simplistic representation and attempted demolition of MacLean's triune brain concept is not good science. Reiner's review, where it has any validity at all, is like discovering a termite or two in the bathroom wall -- and then proceeding to pronounce a full alarm that the house is full of termites -- only to find that it is necessary to treat a couple of boards in the subflooring. Further, in his deconstructive, analytic fervor, Reiner has not offered an alternative higher level generalization. The review represents a dysfunction common to a lot of scientific practice ... that of an analytical approach that takes apart but can't put back together. Perhaps we should call it analytic myopia. Being not interested in the bigger questions of humanity that we so desperately need help on, and lacking an interest in therapy, these analytic myopics continue their fine-grained focus. Fine-grained focus is fine, laudable, and very much needed. It becomes analytically myopic, however, when it fails to place in context what it finds and defines, when it employs sloppy scholarship, and when it attempts prejudicially to destroy or deconstruct that which it lacks the imagination and courage to put together.

On the other hand the theories of brain evolution that Butler and Hodos review favorably and the synthesis which they present at the end of their book focus on the immunohistological, hormonal, and morphological mechanics (1996: 463-73). They say, in fact, almost nothing at all about behavior or the significance for behavioral evolution for the various mechanisms of evolution they identify. And they make no attempt whatsoever to confront the larger behavioral questions of humanity where we need help and guidance from neuroscience in defining the neurobiological basis of human nature in order to establish links up the scale of generalization with the social sciences. The theories they present are only of interest to the technical aspects of neuroscience. They are not, however, incompatible, but rather tend to support MacLean's concepts when these concepts are thoughtfully considered and not inaccurately reported, misrepresented, or grossly oversimplified.

The key point in comparing these theories with that of MacLean's is that they are comparable, at best, only in part. They ask and respond to different questions. MacLean tries to address the larger questions of human nature and behavior. The others show no interest in such questions but address the fine grained technical questions of anatomical and functional evolution. At the level where they meet they do not contradict each other but are largely compatible. At the point they diverge they primarily address different questions. This is, I think, the root of the tension between the two. MacLean's concept facing up the scale of generalization is useful and has been appropriately well-received in the therapeutic sciences, and is also very useful for the social sciences. On the other hand, it has not been, but may yet become, more useful and better received in other quarters of neuroscience ...especially when subjective experience is eventually given its due in the study of consciousness. There are, in fact, recent signs that the importance of subjective experience, which is of great interest to MacLean, is beginning to be more fully recognized in the newer studies of consciousness (Hameroff, et. al. 1996).[141]

The triune brain concept may need modification, then, as the body of neuroscience grows...but certainly not outright rejection. With appropriate clarifications, it is still by far the best concept we have for linking neuroscience with the larger, more highly generalized concepts of the social sciences. This is true even if its level of generality has limited utility for some neuroscience researchers who are doing ever more fine-grained research into neural architecture and function.

[141]See especially the article by Stubenberg (1996); also, Galin (1996: 121) who writes: "I assert that what is most interesting about mental life for most ordinary people is not mechanism, not performance, not information processing; it is what it feels like! Subjective experience!" Searle (1997) provides a general criticism of the emerging consciousness literature. See also the assessment by molecular neuroscientist Smith (1996: 471-74).

The transition from early vertebrate to amniote to synapsid reptile to mammal was in behavioral effect the transition from a nearly exclusively self- preserving organism with relatively little or less complex social life to, at least in part, a nurturing, "other-maintaining", "other-supporting", or "other-interested" organism. And that makes all the difference in the world for human evolution. Our other-maintaining mechanisms combined with our self-preserving ones provide the biological glue as well as the dynamic for our remarkable behavioral evolution, our social life, and ultimately the crucial social and political factor of our moral consciousness.

The qualitative differences between the familial and social behaviors of even the most caring of reptiles (say, modern crocodiles), birds or social insects and the mammal we call human are overwhelmingly evident. Humans with their social, cognitive, and language skills, for better or for worse, dominate the planet and no other species comes close. Any neurobiologist who cannot see or appreciate the difference is suffering from analytic myopia or some form of misplaced species egalitarianism (e.g., see Butler and Hodos 1996: 3-4). The proper study of humans is humans and to some extent their lineal antecedents. The triune brain concept generalizes a fundamental truth out of much that is yet unknown and uncertain in neuroscience. And this generalization, when properly understood, appreciated, and applied, is the most useful bridging link, thus far articulated, between neuroscience and the larger and pressingly critical questions of humanity's survival...as well as the hoped for transformation of humanity into a truly life-supporting, planet-preserving and enhancing custodial species.

When other neuroscience researchers reach the conceptual point in their grasp of the discipline where they feel an increasing obligation to take a more holistic view and proceed to move up the scale of generalization in order to confront the larger questions of human life, they will likely produce concepts closely resembling the triune brain. Homology and behavioral evolution will almost inevitably take them in that direction. When that time comes, if the triune brain concept has been buried in the scrap heap of scientific history, it will be exhumed, refurbished, and honored. Frankly, despite its current lack of popularity in some quarters of neurobiology, I do not think it will be consigned to the scrap heap. I think that it will continue to be influential, and with appropriate modifications as research progresses, provide an important underpinning for interdisciplinary communication and bridging.

Appendix 2

THE PRIMARY ALGORITHMIC LEXICON

The primary [task] *is to show that the apparent richness and diversity of linguistic phenomena is illusory and epiphenomenal, the result of interaction of fixed principles under slightly varying conditions*
(Noam Chomsky 1995: introduction and chapter 4)

The purpose of this appendix is to indicate the correctness of Chomsky's insight quoted above. Chomsky indicated but never achieved a demonstration of the illusory and epiphenomenal nature of the richness and diversity of linguistic phenomena. An analysis of the primary algorithmic basis of both syntax and the lexicon permit substantial progress in achieving this goal. The simplifying effect of the primary algorithm on syntax has been demonstrated in preceding chapters. Here I will focus on the lexicon.

Such a demonstration has been tried before. An earlier attempted simplification of English parts of speech along functional lines was proposed by Fries (1952). He divided words into four major classes named Class 1, 2, 3, & 4. The four classes corresponded functionally to nouns, verbs, adjectives, and adverbs, which make up the bulk of the words in sentences. Fries reported that these four classes, without counting repetitions, accounted for over 93 percent of words in a sample of 1000. All other parts of speech he grouped as "function words" (1952: 65-110). Fries, however, was limited in his theoretical approach because he had no concept of the universal and simplifying algorithmic basis of language and, further, he could not show the basic derivational similarity of many lexical items.

The primary algorithmic syntactic structure allows a further refinement of Fries classification system. This primary algorithmic syntax indicates that only nouns and verbs are basic. Modifiers, such as adjectives and adverbs secondarily attach to the basic categories of nouns and verbs. The function words can be thought of as Fries grouped them. Such a scheme represents a remarkable simplification of linguistic phenomena.

The primary algorithm, however, allows us to go much further in reducing the richness and variety of the actual contents of the lexicon...the words. A remarkable number of words are variants of elements of the primary algorithm itself. Listed below are variants of primary algorithmic terms used in this book. Following that will be other examples from the lexicon of any standard English dictionary. Any reader, by personal effort, may *extend* the list and further *incorporate* a vast number of other lexical items.

1. Variant primary algorithmic terms used throughout the pages of this book: order, incorporation, synthesis, thesis, antithesis, part, whole, unity (primal, cosmic, absolute), equilibrium, equilibration, argument, content, deduction, induction, subject, object, concept, discrimination. Any noun is, of course, an incorporation or an ordering

chaos, extension, process, mulitiplicity, movement, change, reaching, relating, induce, deduce, imbalance. Any verb, of course expresses extension, reach, relation or being.

2. Some synonyms, antonyms, or derivations of primary algorithmic terms: agree, agreement, analyze, analysis arrange, arrangement, assemble, assembly, associate, association, build, building, collect, collection, combine, combination, concept, conception, connect, connection, coordinate, conjunct, conjunction, constitute, constitution, consume, contract, covenant, compact, create, creation, devour, discriminate, eat, forge, form, formation, deformation, gather, gathering, govern, government, group, include, inclusion, induce, institute, institution, introject, join, joining, link, linkage, model, mold, nation, opinion, organ, organism, organize, organization, paradigm, part, pattern, produce, relate, relationship, select, selection, society, structure, swallow, system, systematize, thought, unify, unite, union.

3. A large group of words combine the algorithmic meaning of incorporation and extension. A few illustrative examples are such words as:
appoint, appropriate, assemble, assimilate, augment, capture, clutch, collect, collection colonize, colonialism, control, conquer, coopt, dominate, embezzle, employ, enslave, enthrall, entrance, imprison, inclose, embrace, encarcerate, encircle, encompass, enfold, engage, engorge, enlist, enroll, ensnare, entrap, envelope, expropriate, gather, grasp, group, immobilize, imperialism, include, maintain, obtain, retain, seize, steal, subdue, suborn, subordinate, transcend, transfix, use, utilize

4. Below are words selected from the only first seven pages of the "A" section of Websters New Illustrated Dictionary as they relate to incorporation and extension.
Incorporation:
abandon -- to leave a previous association, attachment, organization, or incorporation
abash -- discompose; disconcert; disorganize, disincorporate

abate -- to reduce; take away; to deduct as a part of a payment or some whole
abdicate -- to renounce a claim, a possession, a thing incorporated to oneself
abhor -- dislike to incorporate, be a part of, or associated with
ablate -- to remove from a whole (incorporation).
abnegate -- deny; withdraw from an incorporation
abolish -- demolish an order or incorporation
abominate -- to strongly reject for incorporation
abort -- to bring to a premature conclusion (incorporation)
abrade -- to scrape away a part of something
abrogate -- demolish something (an incorporation)
absolute -- perfect in order or incorporation
absolve -- to make eligible for re-incorporation
absorb -- assimilate, incorporate
abstain -- to refrain from incorporating
abstemious -- incorporating sparingly
abstinence -- refraining from incorporating
abstract -- separate from an incorporation; detach
abstruse -- hard to mentally incorporate
absurd -- cannot be incorporated; out of fit
abut -- to touch, or join (incorporate)
accept -- to agree to incorporate
access -- admittance to incorporation
accident -- chance disrupting of an event, state, or incorporation
acclaim -- incorporate with honor
acclimate -- to adapt or become incorporated to a new environment.

Extension:
abbreviate -- shorten
abduce -- draw or lead away
abduct -- to carry away (wrongfully)
aberrance -- wandering from straight path
abet -- to aid (an extension) wrongdoing
abide -- to delay extending; stay
ability -- able to extend, and reach out
abjure -- withdraw from an extension; abandon
able -- having power; capable of extending and incorporating
abridge – shorten (an extension)
abroad -- extended away from
abrupt -- halting an extension quickly
abscind -- to cut off an extension
abuse -- to extend and incorporate injuriously

BIBLIOGRAPHY

Adler, M. and W. Gorman. 1952. *The Great Ideas: A Syntopicon of The Great Books of the Western World.* Vol. 1. Encyclopedia Britannica.

Adolphs, R., D. Tranel.,A. Bechara, A. Damasio, H. Damasio. 1996. "Neuropsychological Approaches to Reasoning and Decision-Making." Pp. 157 179 in *Neurobiology of Decision-Making.* Ed. by A.R. Damasio, H. Damasio, and Y. Christen.New York: Springer-Verlag.

Adorno, T., E. Frenkel-Brunswick, D. Levinson, R. Sanford. 1950. *The Authoritarian Personality.* New York: Harper & Brothers.

Aggleton, J. 1992. *The Amygdala: Neurobiological Aspects of Emotion, Memory, and Mental Dysfunction.* New York: Wiley-Liss.

Aldridge, J. Wayne, K. C. Berridge, M. Herman, and L. Zimmer. 1993. "Neuronal Coding of Serial Order: Syntax of Grooming in the Neostriatum." Pp. 391-95 in *Psychological Science.* V. 4, Nr. 6 (Nov).

Allan, S. and P. Gilbert. 1997. "Submissive Behavior and Psychopathology." Pp. 467-488 in *British Journal of Clinical Psychology.* 36

Altman, Joseph and S. A. Bayer. 1997. *Development of the Cerebellar System: In Relation to Its Evolution, Structure, and Functions.* New York: CRC Press

Anderson, James A. 1995. "Associative Networks." Pp. 102-107 in *The Handbook of Brain Theory and Neural Networks.* Ed. by M. Arbib. MIT Press.

Apel, Karl-Otto. 1988. *Diskurs und Verantwortung.* Frankfurt.

Arbib, Michael A. 1995. *The Handbook of Brain Theory and Neural Networks.* MIT Press.

Arens, William. 1998. "Rethinking Anthropophagy." Pp. 39-62 in *Cannibalism and the Colonial World.* Ed. by F. Barker, P. Hulme, and M. Iversen. Cambridge University Press.

Arens, William. 1979. *The Man-Eating Myth: Anthropology and Anthropophagy.* New York: Oxford University Press.

Argyle, M. 1991. *Cooperation: The Basis of Sociality.* Routledge.

Aristotle 1941. *The Basic Works of Aristotle.* Ed. by R. McKeon. New York: Random House.

Arthur, Wallace. 1997. *The Origin of Animal Body Plans: A Study in Evolutionary Developmental Biology.* Cambridge University Press.

Asimov, I. 1989. *Asimov's Chronology of Science and Discovery.* New York: Harper & Row.

Asimov, I. 1962. *Life and Energy*. New York: Doubleday.
Astington, J., P. Harris,. and D. Olson. (eds.) 1988. *Developing Theories of Mind.* Cambridge University Press.
Axley, Milton J. 1998. "Introduction to Peptides and Proteins." Pp. 1-26 in *Bioorganic Chemistry: Peptides and Proteins.* Ed. by S. M. Hecht. Oxford University Press.
Avers, Charlotte J. 1989. *Process and Pattern in Evolution.* Oxford University Press
Baars, Bernard J. 1997. *In the Theatre of Consciousness: The Workspace of the Mind.* Oxford University Press.
Baars, Bernard J. 1988. *A Cognitive Theory of Consciousness.* Cambridge University Press.
Bachevalier, J. and M. Mishkin. 1984. "An early and late developing system for learning and retention in infant monkeys." Pp. 770-778 in *Behavioral Neuroscience* 98.
Baev, Konstantin V. 1998. *Biological Neural Networks: The Hierarchical Concept of Brain Function.* Boston, MA: Birkhauser.
Bailey, Kent. forthcoming. "Upshifting and Downshifting the Triune Brain." In *The Neuroethology of Paul MacLean: Convergences and Frontiers.* Ed. by R. Gardner and G. Cory.
Bailey, Kent. 1987. *Human Paleopsychology.* Hillsdale, N. J.: Lawrence Erlbaum Associates.
Ball, R. C. 1988. "On the Possible Application of Fractal Scaling Ideas in Dendritic Growth." Pp. 239-242 in *Fractals' Physical Origin and Properties.* Ed. by L. Pietronero. New York: Plenum Press.
Barkow, J., L. Cosmides. and J. Tooby. (eds.) 1992. *The Adapted Mind: Evolutionary Psychology and the Generation of Culture.* Oxford University Press.
Barlow, Horace. 1998. "The Nested Networks of Brains and Minds." Pp. 142-155 in *The Limits of Reductionism in Biology.* Ed. by G. R. Bock and J. A. Goode. New York: John Wiley & Sons.
Barnden, John A. 1995. "Artificial Intelligence and Neural Networks."Pp. 102-107 in *The Handbook of Brain Theory and Neural Networks.* Ed. by M. Arbib. MIT Press.
Barnes, R.S.K., P. Calow, and P. Olive. 1993. *The Invertebrates: A New Synthesis.* Oxford: Basil Scientific Publications.
Barnsley, M. 1993. *Fractals Everywhere.* Cambridge: Academic Press.
Bartsch, Renate. 1998. *Dynamic Conceptual Semantics.* Stanford, CA: CSLI Publications.
Baron-Cohen, Simon. 1995. *Mindblindness.* MIT Press
Batson, C. Daniel. 1991. *The Altruism Question: Toward a Social-Psychological Answer.* Hillsdale, New Jersey: Lawrence Erlbaum Associates.
Baudry, M. and J. Davis, (eds.) 1991. *Long-Term Potentiation: A Debate of Current Issues,* MIT Press.
Baynes, Kathleen and J. C. Eliassen. 1998. "The Visual Lexicon: Its Access and Organiztion in Commissurotomy Patients." Pp. 79-104 in *Right Hemisphere Language Comprehension: Perspectives from Cognitive Neuroscience.* Ed. by M. Beeman and C. Chiarello. Mahwah, NJ: Lawrence Erlbaum Associates.
Beardsley, M. C. 1975 *Thinking Straight.* 4th Ed. NJ: Prentice-Hall
Beeman, Mark and Christine Chiarello, (eds.) 1998. *Right Hemisphere Language Comprehension: Perspectives from Cognitive Neuroscience.* Mahwah, NJ: Lawrence Erlbaum Associates.
Bellugi, U., H. Poizner, and E. Klima. 1989. "Language, modality, and the brain." Pp. 380-388 in *Trends in the Neurosciences*, 12.
Bendor, Jonathan and Piotr Swistak. 1997. "The Evolutionary Stability of Cooperation." Pp. 290- 303 in *American Political Science Review.* V.91. N.2
Benecke, R., J. Rothwell, J. Dick, B. Day, and C. Marsden. 1987. "Disturbance of Sequential Movements in Patients with Parkinson's Desease." Pp. 361-80 in *Brain.* V. 110.
Benson, D. 1979. *Aphasia, Alexia, and Agraphia.* Churchill Livingston.
Bergson, H. 1944. *Creative Evolution.* New York: Modern Library.

Berridge, K. C. and J. Fentress. 1988. "Disruption of Natural Grooming Chains after Striatopallidal Lesions." Pp. 336-42 in *Psychobiology*. V. 15.
Berridge, K. C. and I. Whishaw. 1992. "Cortex, Striatum, and Cerebellum: Control of Serial Order in a Grooming Sequence."Pp. 275-90 in *Experimental Brain Research*. V 90.
Bickerton, D. 1995. *Language and Human Behavior*. University of Washington Press.
Blau, Peter M. 1964. *Exchange and Power in Social Life*. New York: John Wiley.
Blessing, William W. 1997. *The Lower Brainstem and Bodily Homeostasis*. Oxford University Press.
Blofeld, J. 1968. *I Ching*. New York: Dutton.
Blum, H. 1968. *Time's Arrow and Evolution*. Princeton University Press.
Blumenfeld, Lev A. and A. Tikhonov. 1994. *Biophysical Thermodynamics of Intracellar Processes*. New York: Springer-Verlag
Bock, Gregory R. and Jamie Goode, (eds). 1998. *The Limits of Reduction in Biology*. Novartis Foundation Symposium 213. New York: John Wiley & Sons.
Bohr, Nils 1987[1934]. *Atomic Physics and the Description of Nature*. Woodbridge, Conn: Ox Bow Press.
Borod, Joan C., R. Bloom, and C. Haywood 1998. "Verbal Aspects of Emotional Communication." Pp. 285-307 in *Right Hemisphere Language Comprehension: Perspectives from Cognitive Neuroscience*. Ed. by M. Beeman and C. Chiarello.. Mahwah, NJ: Lawrence Erlbaum Associates.
Borsley, R. 1991. *Syntactic Theory*. Edward Arnold.
Bowlby, John. 1988. *A Secure Base. Parent-Child Attachment and Healthy Human Development*. New York: Basic Books.
Bowlby, John. 1969. *Attachment*. Vol.1. New York: Basic Books.
Bowlby, John., et. al. 1966. *Maternal Care and Mental Health*. Schocken.
Braine, Martin D. S. 1994. "Is Nativism Sufficient." Pp. 9-31 in *Journal of Child Language* 21.
Braine, Martin D. S. 1993. "The Mental Logic and How to Discover it." in *The Logical Foundations of Cognition*. Ed. by J. Macnamara and G. Reyes. Oxford University Press.
Braine, Martin D. S. 1992. "What Kind of Innate Structure is Needed to 'Bootstrap' into Syntax." Pp. 77-100 in *Cognition*, 45.
Braine, Martin D. S. 1990. "The 'Natural Logic' Approach to Reasoning." Pp. 133-157 in *Reasoning, Necessity, and Logic: Developmental Perspectives*. Ed. by W. Overton. Hillsdale, NJ: Lawrence Erlbaum Associates.
Brandon, R. 1985. "Adaptive explanations: are adaptations for the good of replicators or interactors?" Pp. 81-96 in *Evolution at a Crossroads: The New Biology and the New Philosophy of Science*. Ed. by D. Depew and B. Weber. MIT Press..
Brassard, G. and P. Bratley. 1988. *Algorithmics: Theory and Practice*. NJ: Prentice-Hall.
Bray, Dennis. 1995. "Protein Molecules as Computational Elements in Living Cells." Pp. 307-312 in *Nature*. V. 376. 27 July.
Broda, E. 1975. *The Evolution of Bioenergetic Processes*. Pergamon.
Brooks, D. R. and E. O. Wiley. 1988 *Evolution as Entropy*. 2nd Ed. University of Chicago Press.
Brothers, Leslie. 1995. "Neurophysiology of the Perception of Intentions by Primates." Pp. 1107-1115 in *The Cognitive Neurosciences*. Ed. by M. Gazzaniga. MIT Press.
Brown, Jason. 1977. *Mind, Brain, and Consciousness*. New York: Academic Press.
Brownell, Hiram and Gail Martino. 1998. "Deficits in Inference and Social Cognition: The Effects of Right Hemisphere Brain Damage on Discourse." Pp.309-328 in *Right Hemisphere Language Comprehension: Perspectives from Cognitive Neuroscience*. Ed. by M. Beeman and C. Chiarello. Mahwah, NJ: Lawrence Erlbaum Associates.
Buber, M. 1965. *Between Man and Man*. Trans. by R. Smith. New York: Macmillan.
Buber, M. 1958. *I and Thou*. 2nd Edition. Trans. by R. Smith. New York: Scribner's.
Buchsbaum, R.and L. Milne. 1960. *The Lower Animals*. New York: Doubleday.

Buonomano, D. V. and M. Merznich. 1998. Cortical Plasticity: From Synapses to Maps." Pp. 149-186 in *Annual Review of Neuroscience*. 21.
Burghardt, Gordon M. 1988. "Precocity, Play and the Ectotherm -- Endotherm Transition." Pp.107-48 in *Handbook of Behavioral Neurobiology*, Vol. 9. Ed. by E. M. Bass. New York: Plenum.
Burghardt, Gordon M. 1984. "On the Origins of Play." Pp. 5-41 in *Play in Animals and Humans*. Ed. by P. K. Smith. New York: Basil Blackwell.
Burwick, F. and P. Douglas. 1992. *The Crisis in Modernism: Bergson and the Vitalist Controversy.* Cambridge University Press.
Butler, Ann B. and Hodos, William. 1996. *Comparative Vertebrate Neuroanatomy: Evolution and Adaptation.* New York: Wiley-Liss.
Calvin, William H. 1996. *The Cerebral Code*. MIT Press.
Campbell, B. 1966. *Human Evolution*. Aldine.
Campbell, C. B. G. 1992. "Book Review (MacLean: The Triune Brain in Evolution)". Pp. 497-498 in *American Scientist*, V. 80(Sept-Oct 19 1992).
Caplan, D. 1995. "The cognitive neuroscience of syntactic processing." Pp. 871-879 in *The Cognitive Neurosciences.* Ed. by M. Gazzaniga. MIT Press.
Cariani, Peter. 1994. "As if Time Really Mattered: Temporal Strategies for Neural Coding of Sensory Information." Pp. 209-252 in *Origins: Brain and Self-Organization.* Ed. by K. Pribram. Hillsdale, NJ: Lawrence Erlbaum Associates.
Carruthers, Peter and J. Boucherl. 1998. *Language and Thought: Interdisciplinary Themes.* Cambridge University Press.
Chalmers, David J. 1997. "The Puzzle of Conscious Experience." Pp.30-7 in *Scientific American: Mysteries of the Mind.* Special Issue V. 7, N.1.
Chan, W. 1963. *The Way of Lao Tzu.* Bobs-Merrill.
Chang, C. 1975. *Tao: A New Way of Thinking*. New York: Harper & Row.
Changeux, J-P. and A. Connes. 1995. *Conversations on Mind, Matter, and Mathematics.* Ed. and trans. By M. B. DeBevoise. Princeton University Press.
Changeux, J-P. and S. Dehaene. 1989. "Neuronal models of cognitive functions." Pp. 63-109 in *Cognition*, 33.
Carter, C. Sue, I. Lederhendler, and B. Kirkpatrick. (eds.). 1997. *The Integrative Neurobiology of Affiliation.* New York: Annals of the New York Academy of Sciences (vol. 809).
Cheng, P. W. and K. Holyoak. 1985. "Pragmatic Reasoning Schemas." Pp. 391-416 in *Cognitive Psychology*, 17.
Chester, M. 1993. *Neural Networks: A Tutorial.* Prentice-Hall.
Chiarello, Christine and M. Beeman. 1998. "Introduction to the Cognitive Neuroscience of Right Hemisphere Language Comprehension." Pp. ix-xii. in *Right Hemisphere Language Comprehension: Perspectives from Cognitive Neuroscience*. Ed. by M. Beeman and C. Chiarello. Mahwah, NJ: Lawrence Erlbaum Associates.
Chiari, J. 1992. "Vitalism and contemporary thought," in *The Crisis in Modernism*. Ed. by F. Burwick and P. Douglas. Cambridge University Press.
Chomsky, Noam. 1996. "A Review of B. F. Skinner's Verbal Behavior. Pp. 413-441 in. *Readings in Language and Mind.* Ed. by H. Geirsson and M. Losonsky. Blackwell Publishers.
Chomsky, Noam. 1995. *The Minimalist Program*. MIT Press.
Chomsky, Noam. 1986a. *Barriers*. MIT Press.
Chomsky, Noam. 1986b. *Knowledge of Language: its nature, origin, and use*. New York: Praeger.
Chomsky, Noam.1972. *Mind and Language*. Enlarged edition. New York: Harcourt, Brace, Janovich.
Chomsky, Noam. 1957. *Syntactic Structures*. The Hague: Mouton.
Chong, K. 1990. *Cannibalism in China*. Longwood Academic.

Churchland, Patricia S. and T J. Sejnowski. 1992. *The Computational Brain*. MA: MIT Press.
Churchland, Patricia S. 1986. *Neurophilosophy*. MIT Press.
Cleary, T. 1991. *The Essential Tao*. New York: Harper.
Clutton-Brock, T. H. 1991. *The Evolution of Parental Care*. Princeton University Press.
Coe, Christopher, L. 1990. "Psychobiology of Maternal Behavior in Nonhuman Primates." Pp. 157-83 in *Mammalian Parenting: Biochemical, Neurobiological, and Behavioral Determinants*. Ed. by N. A. Krasnegor and R. Bridges. Oxford University Press.
Corning, Peter A. 1996. "The Cooperative Gene: On the Role of Synergy in Evolution." Pp. 183-207 in *Evolutionary Theory*. V.11.
Corning, Peter A. 1995. "Synergy and Self-organization in the Evolution of Complex Systems.Pp. 89-121 in *Systems Research*. V.12 N.2.
Corning, Peter A. 1983. *The Synergism Hypothesis*. New York: McGraw-Hill.
Cory, Gerald A., Jr. 1999. *The Reciprocal Modular Brain in Economics and Politics: Shaping the Rational Basis of Organization, Exchange and Choice*. New York: Plenum Press.
Cory, Gerald A., Jr. 1998. "MacLean's Triune Brain Concept: in Praise and Appraisal." Pp. 6-19, 22-24. in *Across-Species Comparisons and Psychopathology Society(ASCAP) Newsletter*. V. 11. No. 07.
Cory, Gerald A., Jr. 1997. "The Conflict Systems Behavioral Model and Politics: A Synthesis of Maslow's Hierarchy and MacLean's Triune Brain Concept with Implications for New Political Institutions for a New Century." *Annals of the American Political Science* Association 1997.
Cory, Gerald A., Jr. 1996. *Algorithms, Illusions, and Reality*. Vancouver, WA: Center for Behavioral Ecology.
Cory, Gerald. A., Jr. 1992. *Rescuing Capitalist Free Enterprise for the Twenty First Century*. Vancouver, WA: Center for Behavioral Ecology.
Cory, Gerald A., Jr. 1974. *The Biopsychological Basis of Political Socialization and Political Culture*. Ph.D. Dissertation. Stanford University.
Cosmides, Leda and J. Tooby. 1994. "Better than Rational: Evolutionary Psychology and the Invisible Hand." Pp. 327-32 in *American Economic Review*. V. 84. N.2(May).
Cosmides, Leda and J. Tooby. 1989. "Evolutionary Psychology and the Generation of Culture, Part II." Pp. 51-97 in *Ethology and Sociobiology*. V. 10.
Cowan, W. 1979. "The Development of the Brain." Pp. 56-69 in *The Brain*. Freeman.
Coyle, Joseph T. and S. Hyman. 1999. "The Neuroscientific Foundations of Psychiatry." Pp. 3-33 in *Textbook of Psychiatry*. Ed. by R. Hales, S. Yudofsky, and J. Talbott. Washington, D. C.: American Psychiatric Press.
Crawford, M. 1972. "Chemical evolution." in *Biology of Nutrition*. Ed. by R. T-W-Fiennes. Pergamon.
Crawford, Stephen S. and E. Balon. 1996. "Cause and Effect of Parental Care in Fishes: An Epigenetic Perspective." Pp. 53-107 in *Parental Care: Evolution, Mechanisms, and Adaptive Intelligence*. Ed. by J. Rosenblatt and C. Snowden. New York: Academic Press.
Crews, D. 1997. "Species Diversity and the Evolution of Behavioral Controlling Mechanisms." Pp. 1-21 in *The IntegrativeNeurobiology of Affiliation*. Ed. by C. Carter, I. Lederhendler, and B. Kirkpatrick. New York: Annals of the New York Academy of Sciences (vol. 809).
Crick, Francis. 1994. *The Astonishing Hypothesis: The Scientific Search for the Soul*. New York: Charles Scibner's Sons.
Cromwell, H. C. and K. Berridge. 1990. "Anterior Lesions of the Corpus Striatum Produce a Disruption of Stereotyped Grooming Sequences in the Rat." P. 233 in *Society for Neuroscience* Abstracts. V. 16.

Crump, Martha. L. 1996. "Parental Care among the Amphibia." Pp. 109-144 in *Parental Care: Evolution, Mechanisms, and Adaptive Intelligence.* Ed. by J. Rosenblatt and C. Snowden. New York: Academic Press.

Culicover, Peter W. 1997. *Principles and Parameters: An Introduction to Syntax Theory.* Oxford University Press.

Cummings, Jeffrey L. 1993. "Frontal-Subcortical Circuits and Human Behavior." Pp. 873-880 in *Arch Neurol.*. V. 50(Aug).

Cummins, Denise D. 1998. "Social Norms and Other Minds: The Evolutionary Roots of Higher Cognition." Pp. 30-50 in *The Evolution of Mind.* Ed. by D. Cummins and C. Allen. Oxford University Press.

Damasio, A. R. 1994. *Descarte's Error: Emotion, Reason, and the Human Brain.* New York: Grosset/Putnam

Damasio, H. and Damasio, A. 1980. "The anatomical basis of conduction aphasia." Pp. 337-350 in *Brain,* 103.

Danto, Arthur C. 1965. *Nietzsche as Philosopher.* Columbia University Press.

Darwin, C. 1964. *On the Origin of Species.* A facsimile of the first edition with an intro. by E. Mayr. Harvard University Press.

Daum, Irene and H. Ackermann. 1995. "Cerebellar contributions to cognition." Pp. 201-10 in *Behavioural Brain Research.* 67.

Davidson, R. J. 1995. "Cerebral Asymmetry, Emotion, and Affective style." Pp. 361-387 in *Brain Asymmetry.* Ed. by R. J. Davidson and K. Hugdahl. MIT Press.

Davies, James C. 1991. "Maslow and Theory of Political Development: Getting to Fundmentals." Pp. 389-420 in *Political Psychology.* V. 12. N.3.

Davies, James C. 1963. *Human Nature in Politics.* New York: Wiley.

Davies, P. and J. Gribbin. 1992. *The Matter Myth.* New York: Simon & Schuster.

Davis, G. 1993. *A Survey of Adult Aphasia and Related Language Disorders.* 2nd Edition. Prentice Hall.

Dawkins, R. 1982. *The Extended Phenotype: The Gene as the Unit of Selection.* Freeman.

Dawkins, R. 1976. *The Selfish Gene.* Oxford University Press.

Deacon, Terrence W. 1997. *The Symbolic Species: The Co-evolution of Language and the Brain.* New York: W. W. Norton.

Denbigh, K. and Denbigh, J. 1985. *Entropy in Relation to Incomplete Knowledge.* Cambridge University Press.

Dennett, Daniel C. 1998 "Reflections on Language and the Mind." Pp. 284-294 in *Language and Thought: Interdisciplinary Themes.* Ed. by P. Carruthers and J. Boucher. Cambridge University Press.

Depew, D. and B. Weber. (eds.) 1985. *Evolution at a Crossroads: The New Biology and the New Philosophy of Science.* MIT Press.

Descartes, R. "Rules." *in Great Books of the Western World.* Vol. 31. Ed. by R.M. Hutchins and M.J. Adler. Encyclopedia Britannica.

Devinsky, Orrin and D. Luciano. 1993. "The Contributions of Cingulate Cortex to Human Behavior." Pp. 527-56 in *Neurobiology of Cingulate Cortex and Limbic Thalamus: A Comprehensive Handbook.* Ed. by B. Vogt and M. Gabriel. Boston: Birhauser.

De Waal, Frans B. M. and F. Aurelli. 1997. "Conflict Resolution and Distress Alleviation in Monkeys and Apes." Pp. 317-328 in *The Integrative Neurobiology of Affiliation.* Ed. by C. Carter, I. Lederhendler, and B. Kirkpatrick. New York: Annals of the New York Academy of Sciences (vol. 809).

Dewey, John. 1933. *How We Think.* New York: D. C. Heath.

Dewey, T. G. 1994. "Multifractals and Order-Disorder Transitions in Biopolymers." Pp. 89-100 in *Fractals in the Natural and Applied Sciences.* Ed. by M. Novak. New York: North Holland.

Dewey, T. G. and D. B. Spencer. 1991. "Are Protein Dynamics Fractal?" Pp. 155-171 in *Comments on Molecular and Cellular Biophysics.* V. 7. N.3.

Diamond, A. 1991. "Neuropsychological insights into the meaning of object concept development." Pp. 67-110 in *The Epigenesis of Mind: Essays on Biology and Cognition.* Ed. by S. Carey and R. Gelman. Lawrence Erlbaum.

Dillon, L. 1987. *The Gene: Its Structure, Function, and Evolution.* Plenum Press.

Dimitrov, Mariana; J. Grafman; P. Kosseff; J. Wachs; D. Alway; J. Higgins; I. Litvan; J.S. Lou; and M. Hallett. 1996. "Preserved cognitive processes in cerebellar degeneration." Pp. 131-35 in *Behavioural Brain Research.* 79.

Dingwall, J. 1980. "Human Communicative Behavior: A Biological Model." Pp. 1-86 in *The Signifying Animal.* Ed. by I. Rauch and G. Carr. Indiana University Press.

Donovan, B. 1988. *Humors, Hormones, and the Mind.* Yearbook Medical Publishers.

Dow, R. 1974. "Some novel concepts of cerebellar physiology." Pp. 103-119 in *Mt. Sinai J. Med.* N.Y. 41.

Dowty, D., L. Karttunen, and A. Zwicky, (eds.) 1985. *Natural Language Parsing: Psychological, Computational, and Theoretical Perspectives.* Cambridge University Press.

Dowty, D., R. Wall, and S. Peters. 1981. *Introduction to Montague Semantics.* Reidel..

Driesch, H. 1929. *The Science and Philosophy of the Organism.* 2nd Ed. Black &Co.

Dubnau, J. and T. Tully. 1998. "Gene Discovery in Drosophila: New Insights for Learning and Memory." Pp. 407-444 in *Annual Review of Neuroscience.* V. 21.

Dumitriu, A. 1977. *History of Logic.* Trans. by Abacus staff. Kent: Abacus.

Dyson, F. 1985. *Origins of Life.* Cambridge University Press.

Durfee, E. H. 1993. "Cooperative Distributed Problem-Solving Between (and within) Intelligent Agents." Pp. 84-98 in *Neuroscience: From Neural Networks to Artificial Intelligence.* Ed. by P. Rudomin Heidelberg: Springer-Verlag.

Eccles, J. 1984. "The cerebral neocortex: a theory of its operation." Pp. 1-36 in *Cerebral Cortex, V.2: Functional Properties of Cortical Cells.* Ed. by E. Jones and A. Peters. Plenum Press.

Einstein, A. 1954. "Physics and reality," Pp. 290-323 in *Ideas and Opinions.* Crown.

Eisenberg, Nancy. 1994. "Empathy." Pp. 247-53 in *Encylopedia of Human Behavior.* Ed. by V. S. Ramachandran. New York: Academic Press.

Eckel, Catherine C. and P. Grossman. 1997. "Equity and Fairness in Economic Decisions: Evidence from Bargaining Experiments." Pp.281-301 in *Advances inEconomic Psychology.* Ed. by G. Antonides and W. F. van Raaij. New York: John Wiley & Sons.

Edelman, Gerald M. 1992. *Bright Air, Brilliant Fire.* New York: Basic Books.

Edelman, Gerald M. 1987. *Neural Darwinism: The Theory of Neuronal Group Selection.* New York: Basic Books.

Einstein, Albert. 1954. "Physics and Reality." Pp. 290-323 in *Ideas and Opinions.* New York: Crown Publishing.

Eisenberg, Nancy. 1994. "Empathy." Pp. 247-53 in *Encylopedia of Human Behavior.* Ed. by V. S. Ramachandran. New York: Academic Press.

Ellis, Ralph. 1986 *An Ontology of Consciousness.* Dordrecht, The Netherlands: Martinus Nijhoff.

Elman. J. 1995. "Language processing." in *The Handbook of Brain Theory and Neural Networks.* Ed. by M. Arbib. MIT Press.

Engels, F. 1934. *Dialectics of Nature.* Moscow: Progress Publishers.

Erdal, David and A.Whiten. 1996. "Egalitarianism and Machiavellian Intelligence in Human Evolution." Pp. 139-150 in *Modelling the Early Human Mind.* Ed. by P. Mellars and K. Gibson. Cambridge: The McDonald Institute, University of Cambridge.

Eslinger, Paul. J. 1996. "Conceptualizing, Describing, and Measuring Components of Executive Function." Pp. 367-95 in *Attention, Memory, and Executive Function.* Ed.by G. Lyon and N. Krasnegor. Baltimore: Paul H. Brookes Publishing Co.

Etzioni, Amitai. 1988. *The Moral Dimension.* New York: Free Press.

Fagen, R. 1981. *Animal Play Behavior.* Oxford University Press.

Falmagne, Rachel J. 1990. "Language and the Acquisition of Logical Knowledge." Pp. 111-131 in *Reasoning, Necessity, and Logic: Developmental Perspectives*. Ed. by W. Overton. Hillsdale, NJ: Lawrence Erlbaum Associates.

Fenchel, T. 1987. *Ecology of Protozoa*. Science Tech.

Feynman, R. 1963. *The Feynman Lectures on Physics*. V.2. Addison-Wesley.

Fiengo, R. and R. May. 1996. "Anaphora and Identity." Pp. 117-144 in *The Handbook of Contemporary Semantic Theory*. Ed. by S. Lappin. Blackwell Reference

Fleischer, S. and B. Slotnik. 1978. "Disruption of maternal behavior in rats with lesions of the septal area." Pp. 189-200 in *Physiological Behavior*, 21.

Fleming, Alison S., H. Morgan, and C. Walsh. 1996. "Experiential Factors in Postpartum Regulation of Maternal Care." Pp. 295-332 in *Parental Care: Evolution, Mechanisms, and Adaptive Intelligence*. Ed. by J. Rosenblatt and C. Snowden. New York: Academic Press.

Fleming, A. S., F. Vaccarino, and C. Leubke. 1980 . "Amygdaloid inhibition of maternal behavior in the nulliparous female rat" Pp. 731-743. in *Physiological Behavior*, 25.

Fodor, J. A.. 1983. *The Modularity of Mind*. MIT Press.

Fox, S. 1988. *The Emergence of Life*. New York: Basic Books.

Frank, Robert H. 1988. *Passions Within Reason: The Strategic Role of the Emotions*. New York: W.W. Norton.

Freeman, John H., Jr.; C. Cuppernell; K. Flannery; M. Gabriel. 1996. "Limbic thalamic, cingulate cortical and hippocampal neuronal correlates of discriminative approach learning in rabbits." Pp. 123-36 in *Behavioural Brain Research* 80.

Freud, S. 1962. *Civilization and Its Discontents*. Trans. by J. Strachey. New York: Norton.

Freud, S. 1959. *Group Psychology and the Analysis of the Ego*. Trans. by J. Strachey. New York: Norton.

Fridlund, A. 1991. "Evolution and Facial Action in Reflex, Social Motive, and Paralanguage." Pp. 3-100 in *Biol. Psychol*. 32

Fries, Charles C. 1952. *The Structure of English*. New York: Harcourt, Brace.

Frith, Uta. (1993) 1997. "Autism." Pp. 92-8 in *Scientific American: Mysteries of the Mind*. Special Issue V. 7, N.1.

Frith, Uta. 1989. *Autism: Explaining the Enigma*. Cambridge, MA: Basil Blackwell.

Fromm, E. 1947. *Man for Himself*. Fawcett .

Fuster, Joaquin M. 1997. *The Prefrontal Cortex: Anatomy, Physiology, and Neuro-psychology of the Frontal Lobe*. Third Edition. New York: Lippincott-Raven.

Fuster, J. and R. Bauer. 1974. "Visual short-term memory deficit from hypothermia of frontal cortex." Pp. 393-400 in *Brain Research*. 81.

Gagliardo, Anna; F. Bonadonna; I. Divac. 1996. "Behavioural effects of ablations of the presumed 'prefrontal cortex' or the corticoid in pigeons." Pp. 155-62 in *Behavioural Brain Research*. 78.

Galin, David. 1996. "The Structure of Subjective Experience: Sharpen the Concepts and Terminology." Pp. 121-40 in *Toward a Science of Conciousness: The First Tuscon Discussions and Debates*. Ed, by S. Hameroff, A. Kaszniak; A. Scott. MIT Press.

Galis, F. 1999. "On the Homology of Structures and Hox Genes: the Vertical Column." Pp. 80-91 in *Homology. Novartis Foundation Symposium 222*. Ed. by G. Bock and G Cardew. New York: John Wiley & Sons.

Gans, Carl. 1996. "An Overview of Parental Care among the Reptilia." Pp. 145-57 in *Parental Care: Evolution, Mechanisms, and Adaptive Intelligence*. Ed. by J. Rosenblatt and C. Snowden. New York: Academic Press.

Gardner, Russell, Jr. and Gerald A.Cory, Jr. (eds.) (forthcoming) *The Neuroethology of Paul D. MacLean: Convergences and Frontiers*.

Gardner, Russell, Jr. 1982. "Mechanisms in Manic-Depressive Disorder: An Evolutionary Model." Pp. 1436-1441 *in Archives of General Psychiatry*. 39.

Garrett, M. 1995. "The structure of language processing: Neuropsychological evidence." Pp. 881-899 in *The Cognitive Neurosciences*. Ed. by M. Gazzaniga. MIT Press.
Gayon, J. 1990. "Critics and criticisms of the modern synthesis," Pp. 1-49 in *Evolutionary Biology* Vol. 24. Ed. by M. Hecht, B. Wallace, R. MacIntyre. Plenum Press
Gazdar, G., E. Klein, G. Pullum, and I. Sag. 1985. *Generalized Phrase Structure Grammar*. Basil Blackwell.
Gazzaniga, M.S. 1998. *The Mind's Past*. University of California Press.
Gazzaniga, M.S. 1985. *The Social Brain: Discovering the Networks of the Mind*. New York: Basic Books.
Gedo, J. and A. Goldberg. 1973. *Models of the Mind*. University of Chicago Press.
Gehring, W. J. 1998. *Master Conrol Genes in Development and Evolution*. Yale University Press.
Geirsson, Heimir and M. Losonsky, (eds). 1996. *Readings in Language and Mind*. Blackwell Publishers.
Gell-Mann, M. 1994. *The Quark and the Jaguar*. Freeman.
Geschwind, N. 1965. "Disconnection syndromes in animals and man." Pp. 237-294, 585-644 in *Brain*. 88.
Girault, J-A. and P. Greengard. 1999. "Principles of Signal Transduction." in *Neurobiology of Mental Illness*. Ed. by D. Charney, E. Nestler, and B Bunney. Oxford University Press.
Globus, Gordon, G. 1995. *The Postmodern Brain*. Philadelphia, PA: John Benjamins Publishing Company.
Gloor, Pierre. 1997. *The Temporal Lobe and the Limbic System*. Oxford University Press.
Goldman-Rakic, P. 1987. "Circuitry of primate pre-frontal cortex and regulation of behavior by representational memory." Pp, 373-417 in *Handbook of Physiology*, 5.
Goldstein, M. and I. Goldstein. 1993 *The Refrigerator and the Universe*. Harvard University Press.
Gomez, Juan-Carlos. 1998. "Some Thoughts about the Evolution of LADS, with Special Reference to TOM and SAM." Pp. 76-93 in *Language and Thought: Interdisciplinary Themes*. Ed. by P. Carruthers and J. Boucher. Cambridge University Press.
Goodall, Jane. 1986. *The Chimpanzees of Gombe: Patterns of Behavior*. Harvard University Press.
Gould, S. 1979. *Ever Since Darwin*. New York: Norton.
Greene, Brian 1999 *The Elegant Universe*. New York: W. W. Norton.
Greenough, W., J. Black, and C. Wallace. 1993. "Experience and brain development." Pp. 290-321 in *Brain Development and Cognition*. Ed. by M. Johnson. Blackwell..
Grene M. 1990. "Is evolution at a crossroads?" Pp. 51-81 in *Evolutionary Biology*. Vol. 24 Ed. by M. Hecht, B. Wallace, R. MacIntyre. Plenum Press.
Grice, H. P. 1978. "Further Notes on Logic and Conversation." Pp. in *Syntax and Semantics, IX: Pragmatics*. Ed. by P. Cole. New York: Academic Press.
Grice, H. P. 1975. Logic and Conversation." Pp. in *Syntax and Semantics, III: Speech Acts*. Ed. by P. Cole and J. Morgan. New York: Academic Press.
Groenendijk, J., M. Stokhof, and F. Veltman. 1996. "Coreference and Modality." Pp. 179-213 in *The Handbook of Contemporary Semantic Theory*. Ed. by S. Lappin. Blackwell Reference.
Grossman, J.; A. Carter, F. Volkmar. 1997. "Social Behavior in Autism." Pp. 440-454 in *The Integrative Neurobiology of Affiliation*. Ed. by C. S. Carter, I. Lederhendler and B. Kirkpatrick. New York: Annals of the New York Academy of Sciences (vol. 809).
Grunbaum, A. 1967. *Modern Science and Zeno's Paradoxes*. Middletown, Conn.
Gutnick, M. J. and I. Mody. (eds.) 1995. *The Cortical Neuron*. Oxford University Press.
Habermas, J. 1990 *Justification and Application: Remarks on Discourse Ethics*. Trans. by C. Cronin. MIT Press.
Habermas, J. 1989 *J. Habermas on Society and Politics: A Reader*. Ed. by S. Seidman. Boston: Beacon Press.

Habermas, J. 1987 *Lectures of the Philosophical Discourse of Modernity.* MIT Press.
Halgren, E. 1992. "Emotional Neurophysiology of the Amygdala Within the Context of Human Cognition." Pp. 191-229 in *The Amygdala: Neurobiological Aspects of Emotion, Memory, and Mental Dysfunction.* Ed. by J. Aggleton. New York: Wiley-Liss.
Hall, B.K. 1996. "Homology and Embryonic Development." Pp. 1-37 in *Evolutionary Biology.* V. 28. Ed. by M. K. Hecht, R. J. MacIntyre, and M. T. Clegg. New York: Plenum Press.
Hall, B.K., (ed.) 1994. *Homology: The Hierarchical Basis of Comparative Biology.* San Diego: Academic Press.
Hall, Calvin. 1954. *A Primer of Freudian Psychology.* World Publishing.
Hall, Roland. 1967. "Dialectic," in *The Encyclopedia of Philosophy.* Vol.2. Ed. by P. Edwards. New York: Macmillan.
Hameroff, Stuart R; A. Kaszniak; A. Scott. (eds). 1996. *Toward a Science of Conciousness: The First Tuscon Discussions and Debates.* MIT Press.
Hameroff, Stuart R and R. Penrose. 1996. "Orchestrated Reduction of Quantum Coherence in Brain Microtubules: A Model for Consciousness." Pp. 507-39 in *Toward a Science of Conciousness: The First Tuscon Discussions and Debates.* Ed. by S. Hameroff, A. Kaszniak; A. Scott. MIT Press.
Hamilton, W.D. 1964. "The Genetical Evolution of Social Behavior, I & II." Pp.1-16, &17-52 in *Journal of Theoretical Biology.* V. 7.
Harel, D. 1987. *Algorithms: The Spirit of Compu ting.* New York: Addison-Wesley.
Harlow, Harry F. 1986. *From Learning to Love: The Selected Papers of H. F. Harlow.* Ed by C. Harlow. New York: Praeger.
Harlow, H. 1971. *Learning to Love.* Albion.
Harlow, Harry F and M. K. Harlow. 1965. "The Affectional Systems." Pp. 386-334 in *Behavior of Non-Human Primates.* Ed. by A. M. Schrier, Harry F. Harlow, and F. Stollnitz. New York: Academic Press.
Harrington, Anne, (ed.). 1992. *So Human a Brain.* Boston: Birkhauser.
Harris, M. 1977. *Cannibals and Kings: The Origin of Cultures.* New York: Random House.
Harris, R. 1987. *The Language Machine.* Cornell University Press.
Harth, Erich. 1997. "From Brains to Neural Nets." Pp. 1241-55 in *Neural Networks.* V.10. No.7.
Harth, Erich. 1993. *The Creative Loop.* Reading, MA: Addison-Wesley.
Hartshorne, Charles, Weiss, Paul, and Arthur Burks, (eds.) 1933-58. *Collected Papers of Charles Sanders Peirce.* Harvard University Press.
Havlin, S., S. Buldyrev, A. Goldberger, R. Mantegna, C. Peng, M. Simmons, and H. Stanley, 1995. "Statistical Properties of DNA Sequences." Pp. 1-11 in *Fractal Reviews in the Natural and Applied Sciences.* London: Chapman & Hall.
Hawking, S. 1997. "The Objections of an Unashamed Reductionist." Pp. 169-72 in Roger Penrose. *The Large, the Small, and the Human Mind.* Cambridge University Press.
Haykin, S. 1994. *Neural Networks.* New York: Macmillan.
Hebb, D.O. 1949. *Organization of Behavior: A Neuropsychological Theory.* New York: Wiley.
Hecht, M., B. Wallace, R. MacIntyre, R., (eds.) 1990. *Evolutionary Biology,* Vol. 24, Plenum Press.
Hecht, Sidney M. 1998. *Biorganic Chemistry: Peptides and Proteins.* Oxford University Press.
Heit, G., M. Smith, & E. Halgren. 1988. "Neural encoding of individual words and faces by the hyman hippocampus and amygdala." Pp. 773-775 in *Nature.* 333.
Heisenberg, Werner 1958. "The Representation of Reality in Contemporary Physics." Pp. 95-108 in *Daedelus* 87 (3).

Heller, Wendy; J. Nitschke, and G. Miller. 1998. "Lateralization in Emotion and Emotional Disorders." Pp. 26-32 in *Current Directions in Psychological Science*. Vol. 7. N.1.
Hickman, Cleveland P., Jr., L. Roberts, and F. Hickman. 1990. *Biology of Animals.* Fifth edition. Boston: Times Mirror/Mosby College Publishing.
Hickman, Cleveland P., Jr., L. Roberts, and F. Hickman 1984. *Integrated Principles of Zoology.* Seventh edition. St. Louis: Times Mirror/Mosby College Publishing.
Hinde, R. and Y. Spencer-Booth. 1971. "Effects of a brief separation from mother on rhesus monkeys," Pp. 111-118 in *Science*. Vol. 173.
Ho, M. and P. Saunders, (eds.) 1984. *Beyond Neo- Darwinism.* New York: Academic Press.
Hoffman, M. 1981. "Is Altruism Part of Human Nature?" Pp. 121-37 in *Journal of Personality and Social Psychology*, 40.
Hofstader, Douglas. 1995. *Fluid Concepts and Creative Analogies*. New York: Basic Books.
Hofstader, Douglas. 1989. *Godel, Escher, Bach.* Vintage.
Hogg, G. 1966. *Cannibalism and Human Sacrifice.* Citadel.
Holthoff-Detto, Vjera A.; J. Kessler; K. Herholz; H. Bonner; U. Pietrzyk; M.Wurker; M. Ghaemi; K. Wienhard; R. Wagner; W. Heiss. 1997. "Functional Effects of Striatal Dysfunction in Parkinson Desease." Pp. 145-150 in *Arch Neurol.* V 54(Feb).
Homans, George C. 1961. *Social behavior: Its Elementary Forms*. New York: Harcourt Brace and World.
Homans, George C. 1950. *The Human Group*. New York: Harcourt Brace Jovanovich.
Horn, G. 1993. "Brain mechanisms of memory and predispositions: Interactive studies of cerebral function and behavior." Pp. 481-509 in *Brain Development and Cognition*. Ed. by M. Johnson. Blackwell.
Hornstein, N. 1984. *Logic as Grammar*. Cambridge University Press.
Horrocks, G. 1987. *Generative Grammar*. Longman.
Hull, D. 1981. "The units of evolution," Pp. 23-44 in *Studies in the Concept of Evolut*ion. Ed. by U. Jensen and R. Harre. Harvestor.
Hulme, Peter. 1998. "Introduction: the Cannibal Scene." Pp. 1-38 in *Cannibalism and the Colonial World.* Ed. by F. Barker, P. Hulme, and M. Iversen. Cambridge University Press.
Humphrey, N. K. 1976. "The Function of the Intellect." Pp. 303-17 in *Growing Points in Ethology*. Ed. by P.P.G. Bateson and R. H. Hinde. Cambridge University Press.
Huttenlocher, P. 1993. "Morphometric Study of Human Cerebral Cortex Development." Pp. 112-124 in Brain *Development and Cognition*. Ed. by M. Johnson. Blackwell.
Ingvar, D. 1983. "Serial aspects of language and speech related to prefrontal cortical activity: A selective review." Pp. 177-189 in *Human Neurobiology*. 2.
Inhelder, B. and Piaget, J. 1958. *The Growth of Logical Thinking from Childhood to Adolescence*. New York: Basic Books.
Isaac, Glynn. 1978. "The Food-sharing Behavior of Protohuman Hominids." Pp. 90-108 in *Scientific American*. V.238.
Jackendoff, Ray. 1996. "Semantics and Cognition." Pp. 539-559 in *The Handbook of Contemporary Semantic Theory*. Ed. by S. Lappin. Blackwell Reference.
Jackendoff, Ray. 1994. *Patterns in the Mind: Language in Human Nature*. New York: Basic Books.
Janowsky, J. 1993. "The development and neural basis of memory systems." Pp. 665-678 in *Brain Development and Cognition*. Ed. by M. Johnson. Blackwell.
Janowsky, J., A. Shimamura, and L. Squire. 1989. "Source memory impairment in patients with frontal lobe lesions." Pp. 1043-1056 in *Neuropsychologia.* 27.
Jay, Martin. 1973 *The Dialectical Imagination*. Boston: Little, Brown, and Co.
Jencks, Christopher. 1990 "Varieties of Altruism." Pp. 53-67 in *Beyond Self-Interest*. Ed. by Jane J. Mansbridge. University of Chicago Press.
Jennings, H. 1906. *Behavior of the Lower Organisms*. Columbia University Press.

Johnson, J. and E. Newport. 1989. "Critical period effects in second language learning: The influence of maturational state on the acquisition of English as a second language." Pp. 60-69 in *Cognitive Psychology*. 21.

Johnson, M. 1993. "Brain plasticity: introduction." Pp. 284-289 in *Brain Development and Cognition*. Ed. by M. Johnson. Blackwell.

Johnson-Laird, P. N. 1983. *Mental Models*. Harvard University Press.

Jowett, B., (trans.) 1937. *The Dialogues of Plato*. two vols. New York: Random House.

Jung, Min W. 1995. "Regional Variations in the Hippocampus Place Code." Pp. 65-77 in *Emotion, Memory and Behavior*. Ed. by T. Nakajima and T. Ono. New York: CRC Press.

Kainz, H. 1988. *Paradox, Dialectic, and System*. Penn State University Press.

Kaku, M. 1995. *Beyond Einstein*. Anchor Books.

Kaku, M. 1994. *Hyperspace*. Oxford University Press.

Kalin, N. H. 1997. "The Neurobiology of Fear." Pp.76-83 in *Scientific American Mysteries of the Mind*. Special Issue V. 7, N.1.

Kandel, Eric R., J. H. Schwartz, and T. M. Jessell. 1995. *Essentials of Neural Science and Behavior*. Norwalk, Conn: Appleton & Lange.

Kant, Immanuel. 1969 *Critique of Pure Reason*. Trans. by K. Smith. New York: St. Martin's Press.

Karayiannis, N. and A. Venetsanopoulos (1993). *Artificial Neural Networks*. Kluwer Academic.

Karttunen, L. and A. Zwicky. 1985. "Introduction." Pp. 1-25 in *Natural Language Parsing: Psychological, Computational, and Theoretical Perspectives*. Ed. by D. Dowty, L. Karttunen, and A. Zwicky. Cambridge University Press.

Katz, S., (ed.) 1978. *Mysticism and Philosophical Analysis*. Oxford University Press.

Katz, S., (ed.) 1992. *Mysticism and Language*. Oxford University Press.

Kauffman, S. 1993. *The Origins of Order: Self- Organization and Selection in Evolution*. Oxford University Press.

Kaufmann, W. 1968. *Nietzsche: Philosopher, Psychologist, Antichrist*. 3rd edition. Vintage.

Kempson, Ruth M. 1996. "Semantics, Pragmatics, and Natural-Language Interpretation." Pp. 561-598 in *The Handbook of Contemporary Semantic Theory*. Ed. by S. Lappin. Blackwell Reference.

Kendler, Tracy S. 1995. *Levels of Cognitive Development*. Mahwah, NJ: Lawrence Erlbaum Associates.

Kepnes, S. 1992. *The Text as Thou*. Indiana University Press.

Kilgour, Maggie. 1998. "The Function of Cannibalism at the Present Time." Pp. 238-259 in *Cannibalism and the Colonial World*. Ed. by F. Barker, P. Hulme, and M. Iversen. Cambridge University Press.

Kilgour, Maggie. 1990. *From Communion to Cannibalism: An Anatomy of the Metaphors of Incorporation*. Princeton University Press.

Kimble, D. P., L. Rogers, and C. Hendrickson 1967. "Hippocampal lesions disrupt maternal, not sexual behavior in the albino rat." Pp. 401-05 in *Journal of Comparative Physiological Psychology*. 63.

King, T. and J. Salmon. 1983. *Teilhard and the Unity of Knowledge*. Paulist Press.

Kirk, G. 1954. *Heraclitus, the Cosmic Fragments*. Cambridge University Press.

Kitcher, Philip. 1982. "Genes.", Pp. 337-359 in *British Journal for the Philosophy of Science*. 33..

Kling, A. and L. Brothers. 1992. "The Amygdala and Social Behavior." Pp. 353-377 in *The Amygdala: Neurobiological Aspects of Emotion, Memory, and Mental Dysfunction*. Ed. by J. Aggleton. New York: Wiley-Liss.

Knight, A. 1974. *The Meaning of Teilhard De Chardin*. Devin-Adair.

Knutson, Jeanne N. 1972. *The Human Basis of the Polity*. New York: Aldine-Atherton.

Kohanski, A. 1975. *An Analytical Interpretation of Martin Buber's I and Thou*. Barron's.
Kohlberg, Lawrence. 1984. *The Psychology of Moral Development*. Vol. 2. San Francisco: Harper & Row.
Kohlberg, Lawrence. 1987. *Child Psychology and Childhood Education*. Longman.
Kokman, Emre. 1991. Book Review(MacLean's The Triune Brain: Role in Paleocerebral Functions.) in *J. Neurosurg*. V.75(Dec).
Kolb, B. 1995. *Brain Plasticity and Behavior*." Mahwah, NJ: Lawrence Erlbaum Associates.
Kolb, B. 1993. "Brain Development, Plasticity, and Behavior." Pp. 338-356 in *Brain Development and Cognition*. Ed. by M. Johnson. Blackwell.
Korolev, S., A. Vin'a, V. Tumanyan, and N. Esipova. 1994. "Fractal Properties of DNA Sequences." Pp. 221-228 in *Fractals in the Natural and Applied Sciences*. Ed. by M. Novak. New York: North Holland.
Kosslyn, Stephen M. and O. Koenig. 1995. *Wet Mind: The New Cognitive Neuroscience*. New York: The Free Press.
Krasnegor, Norman A. and R. Bridges. (eds). 1990. *Mammalian Parenting:Biochemical, Neurobiological, and Behavioral Determinants*. Oxford University Press.
Kuntz, P. 1968. *The Concept of Order*. University of Washington Press.
Kutas, M. and S. Hilyard. 1980. "Reading between the lines: Event related potentials during sentence processing." Pp. 354-373 in *Brain Language*. 11.
LaBerge, D. L. 1995. *Attentional Processing: The Brain's Art of Mindfulness*. Harvard University Press.
Lamendella, J. 1977. "The Limbic System in Human Communication." Pp. 157--222 in *Studies in Neurolinguistics*. Vol. 3. Ed. by H. Whitaker and H. A. Whitaker. New York: Academic Press.
Land, Michael F. and R. Fernald. 1992. "The Evolution of Eyes." Pp. 1-29 in *Annu. Rev. Neurosc*. V. 15.
Lashley, K. 1923. "The behavioristic interpretation of consciousness," Pp. 320-353 in *Psychological Review* 30, 1.
LeDoux, Joseph E. 1997 . *The Emotional Brain*. New York: Simon & Schuster.
LeDoux, Joseph E. 1997. "Emotion, Memory, and the Brain." Pp.68-75 in *Scientific American Mysteries of the Mind*. Special Issue V. 7, N.1.
LeDoux, J. 1992. "Emotion and the Amygdala." Pp. 339-351 in *The Amygdala: Neurobiological Aspects of Emotion, Memory, and Mental Dysfunction*. Ed. by J. Aggleton. Wiley-Liss..
Lee, J. 1976. *Patterns of Inner Process*. Citadel Press.
Lehninger, A. 1971. *Bioenergetics*. Benjamin/ Cummings.
Leiner, H., A. Leiner, and R. Dow. 1986. Does the cerebellum contribute to mental skills? Pp. 443-454 in *Behavioral Neuroscience*. 100.
Lenneberg, E. 1967. *Biological Foundations of Language*. New York: Wiley.
Leon, Michael, R. Coopersmith, L. Beasley, and R. Sullivan. 1990. "Thermal Aspects of Parenting." Pp. 400-15 in *Mammalian Parenting: Biochemical, Neurobiological, and Behavioral Determinants*. Ed. by N. A. Krasnegor and R. S. Bridges. Oxford University Press.
Leven, Samuel J. 1994. "Semiotics, Meaning, and Discursive Neural Networks." Pp. 65-82 in *Neural Networks for Knowledge Representation and Inference*. Ed. by D. Levine and M. Aparicio. Hillsdale, NJ: Lawrence Erlbaum Associates.
Levi-Strauss, C [1959] 1969. *The Elementary Structures of Kinship*. London: Eyre and Spottiwoode.
Levin, Beth and M. Hovav. 1996. "Lexical Semantics and Syntactic Structure." Pp. 487-505 in *The Handbook of Contemporary Semantic Theory*. Ed. by S. Lappin. Blackwell Reference.

Levin, F. 1991. *Mapping the Mind.* Analytic Press.
Levine, Daniel S. 1991. *Introduction to Neural and Cognitive Modeling.* Hillsdale, NJ. Lawrence Erlbaum Associates.
Levine, Daniel S. 1986. "A Neural Network Theory of Frontal Lobe Function." Pp, 716--27 In *Proceedings of the Eighth Annual Conference of the Cognitive Science Society.* Hillsdale, NJ: Lawrence Erlbaum Associates.
Levine, Daniel S. and S. Leven. 1992. *Motivation, Emotion, and Goal Direction in Neural Networks.* Hillsdale, NJ. Lawrence Erlbaum Associates.
Levinson, Stephen C. 1995. "Three Levels of Meaning." Pp. 90-115 in *Grammar and Meaning: Essays in Honour of Sir John Lyons.* Ed. by F. Palmer. Cambridge University Press.
Lieberman, Philip. 1998. *Eve Spoke.* New York: Norton.
Lieberman, Philip. 1991. *Uniquely Human: The Evolution of Speech, Thought, and Selfless Behavior.* Harvard University Press.
Linssen, Robert. 1972 *Zen: The Art of Life.* New York: Pyramid.
Llinas, R. 1993. "Coherent 40 Hz Oscillation Characterizes Dream State in Humans." Pp. 2078-2081 in *Proceedings of the National Academy of Sciences.* 90.
Lorenz, Konrad. 1971. *Studies in Animal and Human Behavior.* Vol. 2. Trans. by R. Martin. Harvard University Press.
Lorenz, Konrad. 1970. *Studies in Animal and Human Behavior.* Vol. 1. Trans. by R. Martin. Harvard University Press.
Lorente de No. R. 1949. Pp. 288 315 in *Physiology of the Nervous System.* Ed. by J. Fulton. Oxford University Press.
Losco, Joseph. 1986. "Biology, Moral Conduct, and Policy Science." Pp. 117-44 in *Biology andBureaucracy.* Ed. by E. White and J. Losco. University Press of America.
Lowenberg, J., (ed.) 1957. *Hegel Selections.* New York: Scribner's.
Lowenstein, Werner R. 1999. *The Touchstone of Life.* Oxford University Press.
Lyons, John. 1995. "Grammar and Meaning." Pp. 221-249 in *Grammar and Meaning: Essays in Honour of Sir John Lyons.* Ed. by F. Palmer. Cambridge University Press.
Mac Cormac, E. and M. I. Stamenov. 1996. *Fractals of Brain, Fractals of Mind.* John Benjamin.
MacLean, Paul . D. 1993. "Human Nature: Duality or Triality." Pp. 107-12 in *Politics and the Life Sciences.* V. 12. N.2.
MacLean, Paul . D. 1992. "Obtaining Knowledge of the Subjective Brain("Epistemics")." p.57-70 in *So Human a Brain.* Ed. by Anne Harrington. Boston: Birkhauser.
MacLean, P. D. 1992b. personal letter. dated November 22.
MacLean, Paul D. 1990. *The Triune Brain in Evolution: Role in Paleocerebral Functions.* New York: Plenum Press.
Macphail, Euan M. 1993. *The Neuroscience of Animal Intelligence.* Columbia University Press.
Maddi, Salvatore, R. 1989. *Personality Theories.* 5th Edition. Chicago, Il: The Dorsey Press.
Malinowski, Bronislaw. 1926. *Crime and Custom in Savage Society.* London: Kegan Paul.
Malinowski, Bronislaw. 1922. *Argonauts of the Western Pacific.* London: Routledge & Kegan Paul.
Mansbridge, Jane. (ed.). 1990. *Beyond Self-Interest.* University of Chicago Press.
Mare, C. 1965. "On the degree of generality of the law of negation," *Revista de Filosophie.*
Margolis, Howard. 1982. *Selfishness, Altruism, and Rationality: A Theory of Social Choice.* Cambridge University Press.
Margulis, Lynn. 1998. *Symbiotic Planet: A New Look at Evolution.* New York: Basic Books
Margulis, Lynn and Sagan, Dorion. (1997) *Slanted Truths: Essays on Gaia, Symbiosis and Evolution.* New York: Copernicus Books.
Margulis, Lynn, K. V. Schwartz, and M. Dolan. 1994. *The Illustrated Five Kingdoms.* New York: Harper Collins College Publishers.

Margulis, Lynn and K. V. Schwartz. 1988. *Five Kingdoms*. Second Edition. New York: W.H. Freeman.
Marsden, C.D. 1984. "Which Motor Disorder in Parkinson's Desease Indicates the True Motor Function of the Basal Ganglia. Pp. 225-41 in *Functions of the Basal Ganglia* (Ciba Foundation Symposium 107). London: Pitman.
Marsden, C.D. 1982. "The Mysterious Motor Function of the Basal Ganglia: The Robert Wartenberg Lecture." Pp. 514-39 in *Neurology*. V. 32.
Marx, Karl. 1906. *Capital*. Modern Library.
Maslow, Abraham H. 1968. *Toward a Psychology of Being*. Second Edition. New York: Van Nostrand Reinhold.
Maslow, Abraham H. 1970. *Motivation and Personality*. Second Edition. New York: Harper &Row.
Maslow, Abraham H. 1943. "A Theory of Human Motivation." Pp. 370-96 in *Psychological Review*. V.50.
Masters, Roger D. 1989. *The Nature of Politics*. Yale University Press.
Matthews, Peter. 1995. "Syntax, Semantics, Pragmatics." Pp. 48-60 in *Grammar and Meaning: Essays in Honour of Sir John Lyons*. Ed. by F. Palmer. Cambridge University Press.
Mauss, Marcel. 1954. *The Gift*. Trans. by Ian Cunnison. New York: W.W. Norton.
Maynard Smith, J. 1982. "The Evolution of Social Behavior - a Classification of Models." Pp. 28-44 in *Current Problems in Sociobiology*. Ed by. King's College Sociobiology Group. Cambridge University Press.
McClelland, J. L. and D. Rumelhart, (eds). 1986, *Parallel Distributed Processing: Explorations in the Microstructure of Cognition*. V.2 MIT Press.
McManus, I. and M. Bryden. 1993. "The neurobiology of handedness, language, and cerebral dominance: a model for the molecular genetics of behavior." Pp. 679-701 in *Brain Development and Cognition*. Ed. by M. Johnson. Blackwell.
Mendez-Flor, P. 1989. *From Mysticism to Dialogue: Martin Buber's Transformation of German Social Thought*. Wayne State University Press.
Miller, Edythe S. 1996. "Seen through a Glass Darkly: Competing Views of Equality and Inequality in Economic Thought." Pp. 87-99 in *Inequality: Radical Institutionalist Views on Race, Gender, Class, and Nation*. Ed. by W. M. Dugger. Westport, Connecticut: Greenwood Press.
Miller, Steven A. and J. Harley. 1992. *Zoology*. Duberque, IA: Wm. C. Brown Publishers.
Miller, Trudi C. 1993. "The Duality of Human Nature." Pp. 221-41 in *Poli tics and the Life Sciences*. V. 12. N.2.
Milner, B., M. Petrides, and M. Smith. 1985. Frontal lobes and the temporal organization of memory." Pp. 137-142 in *Human Neurobiology*. 4.
Minsky, M. 1979. "The Social Theory". Pp. 423-50 in *Artificial Intelligence: an MIT Perspective*. Vol. 1. Ed. by P. Winston and R. Brown. Cambridge, MA: MIT Press.
Mirsky, Allan F. 1996. "Disorders of Attention: A Neuropsychological Perspective." Pp. 71-95 in *Attention, Memory, and Executive Function*. Ed. G. Lyon and N.Krasnegor Baltimore: Paul H. Brookes Publishing Co.
Montague, R. 1970. "Universal Grammar." Pp. 373-398 in *Theoria* 36.
Morowitz, H. 1985. *Mayonnaise and the Origin of Life*. New York: Scribner's.
Morowitz, H. 1978. *Foundations of Bioenergetics*. New York: Academic Press.
Moser, Paul K. 1993. *Philosophy after Objectivity*. Oxford University Press.
Motz, L. 1987. "Introduction" to the *World of Physics*. V. I. Ed. by J. H. Weaver. New York: Simon and Schuster.
Mountcastle, V. 1978. "An organizing principle for cerebral function." Pp. 5-50 *in The Mindful Brain*. Ed. by Mountcastle and G. Edelman. MIT Press.

Murdoch, B., R. Afford, A, Ling, and B. Ganguley 1986. "Acute computerized tomographic scans: Their value in the localization of lesions and as prognostic indicators in aphasia." Pp. 311-345 in *Journal of Communication Disorders*, 19.

Nagai, Katsuya and Hachiro Nakagawa. 1992 *Central Regulation of Energy Metabolism with Special Reference to Circadian Rhythm.* Tokyo: CRC Press.

Naranjo, C. and R. Ornstein. 1971. *The Psychology of Meditation.* Shambala.

Nestler, E. J. and S. E. Hyman 1999. "Mechanisms of Neural Plasticity." Pp. 61-72 in *Neurobiology of Mental Illness.* Ed. by D. Charney, E. Nestler, and B Bunney. Oxford University Press.

Newman, James; B. Baars, and Sung-Bae Cho. 1997. "A Neural Global Workspace Model for Conscious Attention." Pp. 1195-1206 in *Neural Networks.* Vol 10. No. 7.

Newton, Natika. 1996. *Foundations of Understanding.* Amsterdam/Philadelphia: John Benjamins Publishing Co.

Nicolis, G. and Prigogine, I. 1977. *Self-Organization in Nonequilibrium Systems.* New York: Wiley.

Neville, H. 1993. "Neurobiology of cognitive and language processing: effects of early experience." Pp. 424-448 in *Brain Development and Cognition.* Ed. by M. Johnson. Blackwell..

Neville, H.; J. Nicol; A. Barss; K. Forster; and M. Garrett. 1991. "Syntactically based sentence processing classes: Evidence from event related potentials." Pp. 151-165 in *Journal of Cognitive Neuroscience* 3.

Newport, E. and T. Suppala. 1987. *A Critical Period Effect in the Acquisition of a Primary Language.* University of Illinois Press.

Nielsen, Claus. 1995 *Animal Evolution: Interrelationships of the Living Phyla.* Oxford University Press.

Nietzsche, F. 1967/1900. *The Will to Power.* Trans. by W. Kaufmann and R.J. Hollingdale. New York: Random House.

Niwa, S. 1989. "Schizophrenic symtoms, pathogenic cognitive and behavioral features: Discussion of the 'language of brain' and of 'mind'." Pp. 83-91 *in Main Currents in Schizophrenia Research.* Ed. by M. Namba and H. Kaiya. Tokyo: Hesco International(in Japanese, reported in Levin 1991).

Novak, M. M. (ed.) 1995. *Fractal Reviews in the Natural and Applied Sciences.* London: Chapman & Hall.

Novak, M. M. (ed.) 1994. *Fractals in the Natural and Applied Sciences.* New York: North Holland.

Numan, Michael and T. Sheehan. 1997. "Neuroatomical Circuitry for Mammalian Maternal Behavior." Pp. 101-125 in *The Integrative Neurobiology of Affiliation.* Ed. by C. Carter, I. Lederhendler, and B. Kirkpatrick. New York: Annals of the New York Academy of Sciences (vol. 809).

Numan, Michael. 1994. "Maternal Behavior." Pp. 221-302 in *Physiology of Reproduction.* 2nd edition. Vol. 2. Ed. by E. Knobil and J. Neill. New York: Raven Press.

Numan, Michael. 1990. "Neural Control of Maternal Behavior," Pp. 231-59 in *Mammalian Parenting: Biochemical, Neurobiological, and Behavioral Determinants.* Ed. by N. A. Krasnegor and R. S. Bridges. : Oxford University Press.

Numan, M., M. J. Numan, and J. English. 1993. "Excito-toxic amino acid injections into the medial amygdala facilitate maternal behavior in virgin female rats. Pp. 56-81 in *Horm. Behav.* 27.

Oberg, R.G.E. and I. Divac. 1979. "'Cognitive' functions of the Neostriatum." Pp. 291-313 in *The Neostriatum.* Ed. by I. Divac and R.G.E. Oberg. Oxford: Pergamon Press.

O'Keefe, J. and J. Dostrovsky. 1971. Pp. 171-175 in *Brain Research* 34.

O'Leary, D. 1989. "Do cortical areas emerge from a protocortex?" Pp. 400-406 in *Trends in Neuroscience* 12 (Elsevier Trends Journals).

Ono, T., S. Eifuku, H. Nishijo, and R. Tamura. 1995. "Place-Correlates of Monkey Hippocampal Neuronal Activity and its Significance in Memory." Pp. 139-152 in *Emotion, Memory and Behavior*. Ed. by T. Nakajima and T. Ono. New York: CRC Press.

Oparin, A.I. 1964. *Life: Its Nature, Origin, and Development*. New York: Academic Press.

Otto, R. 1975. *Mysticism East and West*. Meridian.

Overton, Willis F. 1990. *Reasoning, Necessity, and Logic: Developmental Perspectives*. Hillsdale, NJ: Lawrence Erlbaum Associates.

Palmquist, Stephen 1995. *The Tree of Philosophy*. Hong Kong: Philopsychy Press.

Panksepp, Jaak. 1998. *Affective Neuroscience*. Oxford University Press.

Parent, A. 1986. *Comparative Neurobiology of the Basal Ganglia*. New York: Wiley.

Penrose, Roger 1997. *The Large, the Small, and the Human Mind*. Cambridge University Press.

Penrose, Roger. 1994. *Shadows of the Mind*. Oxford University Press.

Penrose, Roger. 1991. *The Emperor's New Mind*. Penguin.

Perls, F. S. 1947. *Ego, Hunger, and Aggression*. New York: Random House.

Perrett, D., M. Oram, E. Wachsmuth, and N. Emery. 1995. "Understanding the Behavior and 'Minds' of Others from their Facial and Body Signals: Studies of Visual Processing within the Temporal Cortex." Pp. 155-167 in *Emotion, Memory and Behavior*. Ed. by T. Nakajima and T. Ono. New York: CRC Press.

Perrett, D., A. Mistlin, & A. Chitty. 1987. "Visual neurons responsive to faces." Pp. 358-364 in *Trends in Neuroscience*, 10.

Petitto, Laura A. 1997. "In the Beginning: On the Genetic and Environmental Factors That Make Early Language Acquisition Possible." Pp. 45-69 in *The Inheritance and Innateness of Grammars*. Ed. by M. Gopnik. Oxford University Press 1997.

Phillips, Jerry. 1998. "Cannibalism qua Capitalism: the Metaphorics of Accumulation in Marx, Conrad, Shakespeare and Marlow. Pp. 183-203 in *Cannibalism and the Colonial World*. Ed. by F. Barker, P. Hulme, and M. Iversen. Cambridge University Press.

Piaget, Jean. 1977. *The Development of Thought*. Trans. by A. Rosen. New York: Viking Press.

Piaget, Jean. 1970. "Piaget's Theory" in *Carmichael's Manual of Child Psychology*. Ed. by P. H. Mussen. New York: Wiley.

Piaget, Jean. [1947] 1968. *Psychology of Intelligence*. Littlefield, Adams.

Piaget, Jean. 1953. *Logic and Psychology*. Manchester University Press.

Piaget, Jean. [1932] 1965. *The Moral Judgement of the Child*. New York: Free Press.

Pickover, C. A. 1996. *Fractal Horizons*. New York: St. Martin's Press.

Pietroner, Luciano.(ed.) 1988. *Fractals' Physical Origin and Properties*. New York: Plenum Press.

Pinker, Steven. 1997. *How the Mind Works*. New York: Norton.

Pinker, Steven. 1995. "Introduction"(to language). Pp. 851-853 in *The Cognitive Neurosciences*. Ed. by M. Gazzaniga. MIT Press.

Pinker, Steven. 1994. *The Language Instinct*. New York: Morrow.

Planck, M. 1987. "Dynamical laws and statistical laws." Pp. 88-100 in *The World of Physics*. Vol. 1. Ed. by J. Weaver. New York: Simon and Schuster.

Plato 1937. *The Dialogues of Plato*. two vols. Trans. by B. Jowett. New York: Random House.

Ploog, Detlev W. 1992. "Neuroethological Perspectives on the Human Brain: from the Expression of Emotions to Intentional Signaling." Pp. 3-13 in *So Human a Brain*. Ed. by Anne Harrington. Boston: Birkhauser.

Poizner, H., E. Klima, and U. Bellugi. 1987. *What the Hands Reveal about the Brain*. MIT Press.

Pooley, A.C. 1977. "Nest opening response of the Nile crocodile." Pp. 17-26 in *Journal of Zoology*(London), 182.

Posner, M. and M. Raichle. 1994. *Images of the Mind*. Scientific American Library.

Powell, T. 1981. Pp 1-19 in *Brain Mechanisms and Perceptual Awareness*. Ed. by O. Pompeiano and A. Marsan, pp. 1-19. Raven Press.

Premack, D. and G. Woodruff. 1978. "Does the Chimpanzee have a Theory of Mind?" Pp. 515-526 in *Behavioral Brain Science*. 1.

Pribram, Karl H. 1998 "Inaugural Lecture." *Meeting of the Society for the Multidisciplinary Study of Consciousness,* San Francisco, CA.

Pribram, Karl H. 1994. "Brain and the Structure of Narrative." Pp. 375-415 in *Neural Networks for KnowledgeRepresentation and Inference*. Ed. by D. Levine and M. Aparicio. Hillsdale, NJ: Lawrence Erlbaum Associates.

Pribram, Karl H. 1991. *Brain and Perception: Holonomy and Structure in Figural Processing.* Hillsdale, NJ: Lawrence Erlbaum Associates.

Pribram, Karl H. 1973. "The Primate Frontal Cortex --Executive of the Brain." Pp. 293-314 in *Psychophysiology of the Frontal Lobes*. Ed. by K. Pribram and A. Luria. New York: Academic Press.

Price, John S., L. Sloman, R. Gardner, P. Gilbert, and P. Rohde. 1994. "The Social Competition Hypothesis of Depression." Pp. 309-315 in *British Journal of Psychiatry*. 164.

Price, John S. and, Leon Sloman. 1987. "Depression as Yielding Behavior: An Animal Model Based of Schjelderup-Ebbe's Pecking Order." Pp. 85S-98S in *Ethology and Sociobiology*. 8.

Price, John S. 1967. "Hypothesis: The Dominance Hierarchy and the Evolution of Mental Illness." Pp. 243-246 in *Lancet* 11

Prigogine, Ilya. 1994. "Mind and Matter: Beyond the Cartesian Dualism." Pp. 2-15 in *Origins: Brain and Self-Organization.* Ed. by Karl Pribram. Hillsdale, NJ: Lawrence Erlbaum Associates.

Prigogine, Ilya. 1983. "Time and the Unity of Knowledge," in *Teilhard and the Unity of Knowledge.* Ed. by T. King, and J. Salmon. Paulist Press.

Prigogine, Ilya. 1980. *From Being to Becoming*. Freeman.

Prigogine, Ilya. and Allen, P. 1982. "The challenge of complexity," Pp. 3-38 in *Self-Organization and Dissipative Structures*. Ed. by W. Schrieve and P. Allen. University of Texas Press.

Prigogine, Ilya and I. Stengers. 1984. *Order Out of Chaos: Man's New Dialogue with Nature.* New Science Library.

Putnam, Hilary 1983. *Realism and Reason* (Philosophical Papers, Volume 3). Cambridge University Press.

Quinn, W. G. 1998. "Reductionism in Learning and Memory." Pp. 117-132 in *The Limits of Reductionism in Biology.* Ed. by G. R. Bock and J. A. Goode. New York: John Wiley& Sons.

Radford, Andrew. 1997. *Syntax: A Minimalist Introduction.* Cambridge University Press.

Rajchman, John and West, Cornel, (ed.) 1985. *Post-Analytic Philosophy.* Columbia University Press.

Reiner, Anton, 1990. "An Explanation of Behavior"(review of MacLean's The Triune Brain in Evolution). Pp. 303-05 in *Science*, V. 250(Oct 12, 1990).

Rensch, B. 1971. *Biophilosophy.* Trans. C. Sym. Columbia University Press.

Restak, Richard M. 1994. *The Modular Brain.* A Lisa Drew Book. New York: Charles Scribner's Sons.

Romer, A. 1958. "Phylogeny and behavior with special reference to vertebrate evolution." In *Behavior and Evolution.* Ed. by A. Roe and G. Simpson. Yale University Press.

Rorty, R. 1985. "Solidarity or Objectivity?" in *Post-Analytic Philosophy.* Ed. by J. Rajchman and C. West. Columbia University Press.

Rorty, R. 1979. *Philosophy and the Mirror of Nature*. Princeton University Press.

Rosenblatt, Jay S. and C. Snowden. 1996. *Parental Care: Evolution, Mechanisms, and Adaptive Intelligence.* New York: Academic Press.

Roth, V. Louise. 1994. "Within and Between Organisms: Replicators, Lineages, and Homologues." Pp. 301-37 in *Homology: The Hierarchical Basis of Comparative Biology*. Ed. by B. K. Hall. New York: Academic Press.
Russell-Hunter, W. 1969. *A Biology of Higher Invertebrates*. New York: Macmillan.
Rutkevich, M. 1958. "The nature of the law of negation of the negation and its sphere of action." in *Filosofskie Nauki*. Moscow.
Ryle, G. 1954. *Dilemmas*. Cambridge University Press.
Sacks, O. 1989. *Seeing Voices*. University of California Press.
Sagan, Carl. 1977. *The Dragons of Eden*. New York: Random House.
Sagan, E. 1974. *Cannibalism*. New York: Harper & Row.
Sahlins, Marshall. 1976. *Culture and Practical Reason*. University of Chicago Press.
Sahlins, Marshall. 1972. *Stone Age Economics*. Chicago, IL: Aldine-Atherton..
Sahlins, Marshall. 1963. "On the Sociology of Primitive Exchange." Pp. 139-236 *in The Relevance of Models for Social Anthropology*. Ed. by Michael Banton. London: Tavistock Publications.
Salter, Frank K. 1995. *Emotions in Command: A Naturalistic Study of Institutional Dominance*. Oxford University Press.
Salthe, G. 1972. *Evolutionary Biology*. New York: Holt, Rinehart &Winston.
Sanday, P. 1986. *Divine Hunger: Cannibalism as a Cultural System*. Cambridge University Press.
Sarter, M. and H. Markowitsch. 1985. "Involvement of the Amygdala in learning and memory: a critical review, with emphasis on anatomical relations." Pp. 342-380 in *Behavioral Neuroscience*. 99.
Sartre, J. 1976. *Critique of Dialectical Reason*. Trans. by A. Smith. Humanities Press.
Sartre, J. 1963. *The Problem of Method*. Trans. by H. Barnes. Methuen.
Saver, J. L. and J. Rabin. 1997. "The Neural Substrates of Religious Experience." Pp. 195-207 in *The Neuropsychiatry of Limbic and Subcortical Disorders*. Ed. by S. Salloway, P. Malloy, and J. Cummings. Washington, D.C.: American Psychiatric Press.
Sayre, K. 1969. Plato's Analytic Method. University of Chicago Press.
Scanlan, J. 1985. *Marxism in the USSR*. Cornell University Press.
Schein, Edgar H. 1961. *Coercive Persuasion*. New York: W.W. Norton.
Schiller, P. 1985. "A model for the generation of visually guided sacadic eye movements." in *Models of the Visual Cortex*. Ed, by D. Rose and V. Dodson. New York:. Wiley.
Schmajuk, N. A. 1997. *Animal Learning and Cognition: A Neural Network Approach*. Cambridge University Press.
Schrieve, W. and P. Allen. 1982. *Self-Organization and Dissipative Structures*. University of Texas Press.
Schroedinger, E. 1944. *What is Life?* Cambridge Edition (1967). Cambridge University Press.
Schroedinger, E. 1958. *Mind and Matter*. Cambridge University Press.
Schopf, J. W. 1996. "Are the Oldest Fossils Cyanobacteria?" Pp. 23-61 in *Evolution of Microbial Life*. Ed. by D. Roberts. Cambridge University Press.
Schopf, J. W. 1993. "Microfossils of the Early Archean Apex Chert: New Evidence of the Antiquity of Life. Pp. 640-646 in *Science*, 260.
Searle, J. R. 1997. *The Mystery of Consciousness*. The New York Times Review of Books.
Simmel, G. 1950. *The Sociology of Georg Simmel*. Trans. by K. Wolff. Free Press.
Sinnott, E. 1950. *Cell and Psyche*. University of North Carolina Press.
Siu, R. 1968. *The Man of Many Qualities: A Legacy of the I Ching*. MA: MIT Press.
Skinner, B. F. 1971. *Beyond Freedom and Dignity*. New York: Knopf.
Skinner, B. F, 1957. *Verbal Behavior*. New York: Appleton- Century-Crofts.
Skinner, B. F. 1948. *Walden Two*. New York: Macmillan Paperback(1962).
Skinner, B. F. 1938. *The Behavior of Organisms*. New York: Appleton-Century-Crofts

Slotnik, B. M. 1967. "Disturbances of maternal behavior in the rat following lesions of the cingulate cortex." Pp. 204-36 in *Behavior*, 29.

Smith, Adam. 1977[1740-90]. "The Correspondence of Adam Smith,(1740-90)", Mossner, E. C.and Ross, T. S. Eds., in Vol. 6 of *The Glasgow Edition of the Works and Correspondence of Adam Smith.* General editing by D. D Raphael and A. Skinner. Oxford: Clarendon Press.

Smith, Adam. 1911[1789]. *The Theory of Moral Sentiments*. New edition. London: G. Bell.

Smith, Adam. 1937[1776]. *The Wealth of Nations*. Ed. by Edwin Cannan. New York: Modern Library.

Smith, C. A. and E. J. Wood. 1992. *Biosynthesis*. London: Chapman & Hall.

Smith, C.U.M. (forthcoming). "Deep Time and the Brain: the Message of the Molecules." In *The Neuroethology of Paul D. MacLean: Convergences and Frontiers*. Ed. by R. Gardner, Jr. and G. Cory, Jr.

Smith, C.U.M. 1996 *Elements of Molecular Neurobiology*. 2nd Edition New York: John Wiley & Sons.

Smith, K., Jr.; P. Bol.; J. Adler; & D. Wyatt. 1990. *Sung Dynasty Uses of the I Ching*. Princeton University Press.

Smith, M. Brewster. 1991. "Comments of Davies's 'Maslow and Theory of Political Development." Pp. 421-23 in *Political Psychology*. V.12. N.3.

Smith, R. 1958. *I and Thou, by Martin Buber*. New York: Scribners.

Smuts, J. C. 1926. *Holism and Evolution*. New York: Macmillan.

Snow, C. 1977. "Mothers' speech research: From input to interaction." Pp. 31-49 in *Talking to Children: Language input and interaction*. Ed. by C. Snow and C. Ferguson. Cambridge University Press.

Somit, Albert and Steven A. Peterson. 1997. *Darwinism, Dominance, and Democracy*. Westport, Conn: Praeger Publishers.

Sommers, F. T. 1982 *The Logic of Natural Language*. Oxford University Press.

Spitz, Rene A. 1965. *The First Year of Life*. New York: International Universities Press.

Spitz, Rene A. 1945. "Hospitalism: an inquiry into the genesis of psychiatric conditions in early childhood." in *The Psychoanalytic Study of The Child.* Vol.1. Ed by O. Fenichel. International Universities.

Staal, F. 1975. *Exploring Mysticism*.University of California Press.

Stace, W. 1960. *Mysticism and Philosophy*. Lippincott.

Stamm, J. S. 1955. "The function of the medial cerebral cortex in maternal behavior in rats." Pp. 347-56 in *Journal of Comparative Physiol. Psychol.* 48.

Stapp, Henry P. 1972. "The Copenhagen Interpretation." Pp. 1098-1116 in *American Journal of Physics*. 40(8).

Steiner, Kurt. (forthcoming). *The Tokyo Trials*.

Stern, Chantal E. and R. Passingham. 1996. "The nucleus accumbens in monkeys (Macaca fascicularis): II. Emotion and motivation." Pp. 179-93 in *Behavioral Brain Research*. 75.

Stevens, C. 1996 "Spatial Learning and Memory: The Beginning of a Dream." Pp. 1147-1148 in *Cell*. 87.

Strickberger, Monroe W. 1996. *Evolution*. Second Edition. Boston: Jones and Bartlett.

Striedter, Georg F. 1997. "The Telencephalon of Tetrapods in Evolution." Pp. 179-213 in *Brain, Behavior, and Evolution*. 49.

Stromswold, K. 1995. "The cognitive and neural bases of language acquisition." Pp. 855-870 in *The Cognitive Neurosciences*. Ed. by M. Gazzaniga. MIT Press.

Strother, Paul. 1992. "Evidence of Earliest Life." Pp. 87-101 in *Environmental Evolution*. Ed. by L. Margulis and L. Olendzenski. The MIT Press.

Strum, Shirley S. and B. Latour. 1991. "Redefining the Social Link From Baboons to Humans." Pp 73-85 in *Primate Politics*, edited. by G. Schubert and R. D. Masters. Southern Illinois University Press.

Stubenberg, Leopold. 1996. "The Place of Qualia in the World of Science." Pp. 41-9 in *Toward a Science of Conciousness: The First Tuscon Discussions and Debates*. Ed, by S. Hameroff, A. Kaszniak, A. Scott. MIT Press.

Stuss, D. and D. Benson. 1986. *The Frontal Lobes*. Raven Press.

Suzuki, D. 1972. *What is Zen?* New York: Harper & Row.

Tarski, A. 1944. "The Semantic Conception of Truth."Pp. 341-375 in, *Philosophy and Phenomenological Research*. V. 4.

Taylor, Talbot J. 1997. *Theorizing Language: Analysis, Normativity, Rhetoric, History*. New York: Pergamon.

Teilhard de Chardin, P. 1973. *Man's Place in Nature*. Trans. by R. Hague. Harper.

Teilhard de Chardin, P. 1971. *Activation of Energy*. Trans. by R. Hague. New York: Harcourt, Brace, Jovanovich.

Teilhard de Chardin, P. 1969. *Human Energy*. Trans. by J. Cohen. New York: Harcourt Brace Jovanovich.

Teilhard de Chardin, P. 1965. *The Phenomenon of Man*. Harper.

Terlecki, L.J. and R. Sainsbury. 1978. "Effects of fimbria lesions on maternal behavior of the rat." Pp. 89-97 in *Physiological Behavior*, 21.

Teske, Nathan. 1997. "Beyond Altruism: Identity-Construction as Moral Motive in Political Explanation." Pp. 71-91 in *Political Psychology*. V.18. N.1.

Thelen, E. 1989. "Self-organization in developmental processes: Can systems approaches work?" Pp. 77-117 in *Systems and Development. The Minnesota Symposium in Child Psychology*. Vol. 22. Ed. by M. Gunnar and E. Thelen. Lawrence Erlbaum.

Thomson, K. S. 1988. *Morphogenesis and Evolution*. Oxford University Press.

Tipler, Frank. 1994. *The Physics of Immortality*. New York: Doubleday.

Tonomi, Giulio and G. Edelman. 1998. "Consciousness and Complexity." Pp. 1846-1851 in *Science*. Vol 282(4 Dec)

Tooby, John and L. Cosmides. 1989. "Evolutionary Psychology and the Generation of Culture, Part I." Pp. 29-49 in *Ethology and Sociobiology*. V. 10.

Tooby, John and I. DeVore. 1987 "The Reconstruction of Hominid Behavioral Evolution through Strategic Modeling." in *Primate Primate Models for the Origin of Human Behavior*. Ed. by W. G. Kinsey. SUNY Press.

Trimble, M. R. 1996. *Biological Psychiatry*. 2nd Edition. New York: John Wiley.

Trivers, R.L. 1981. "*Sociobiology and Politics*." Pp. 1-44 in *Sociobiology and Human Politics*, Ed. by E. White. Lexington, Mass.: D.C. Heath & Company.

Trivers, R.L. 1971. "The Evolution of Reciprocal Altruism." Pp. 35-57 in *The Quarterly Review of Biology*. V. 46.

Trumbo, Stephen T. 1996. "Parental Care in Invertebrates," Pp.3-51 in *Parental Care: Evolution, Mechanisms, and Adaptive Intelligence*. Ed. by J. Rosenblatt. and C. Snowden, New York: Academic Press.

Tsunoda, T. 1987. *The Japanese Brain*. Tokyo: Daishu Shoten (in Japanese, reported in Levin 1991).

Tucker, Don M.; P. Luu, Phan; and K. Pribram. 1995. "Social and Emotional Self-Regulation." Pp. 213-39 in *Annals of the New York Academy of Sciences*. Vol. 769.

T-W-Fiennes, R. (ed.) 1972. *Biology of Nutrition*. Pergamon.

Underhill, Evelyn. 1961. *Mysticism*. Dutton.

Uriagereka, J. 1998. *Rhyme and Reason: An Introduction to Minimalist Syntax*. MIT Press.

Van Loocke, Philip. 1999. "Complex Systems Methods in Cognitive Systems and the Representation of Environmental Information." Pp. 91-114 in *The Nature of Concepts*. Ed. by P Van Loocke. London and New York: Routledge.

Van Over, Raymond. 1973 *Chinese Mystics*. New York: Harper & Row.

Van Valen, Leigh, M. 1982. "Homology and Causes." Pp. 305-12 in *Morphology* 173.

Van Valin, Robert D., Jr. and R. LaPolla. 1997. *Syntax: Structure, Meaning and Function*. Cambridge University Press.

Vandervert, L. 1990. "Systems thinking and neurological positivism: further elucidations and implications." Pp. 1-17 in *Systems Res.* 7.

Vandervert, L. 1988. "Systems thinking and a proposal for a neurological positivism." Pp. 313-321 in *Systems Res.* 5.

Veenman, C. Leo, L. Medina, and A. Reiner. 1997. "Avian Homologues of Mammalian Intralaminar, Mediodorsal and Midline Thalamic Nuclei: Immunohistochemical and Hodological Evidence." Pp. 78-98 in *Brain, Behavior, and Evolution.* 49.

Waal, Frans de. 1996. *Good Natured: The Origins of Right and Wrong in Humans and Other Animals.* Harvard University Press.

Wallace, R. A. and A. Wolf. 1991 *Contemporary Sociological Theory: Continuing the Classical Tradition.* Prentice-Hall.

Waltz, J.; B. Knowlton.; K. Holyoak; K. Boone; F. Mishkin; M. de Menezes Santos; C. Thomas; and B. Miller. 1999. "A System for Relational Reasoning in Human Prefrontal Cortex." Pp. 119-128 in *Psychological Science.* V. 10. N. 2.

Warren, S. 1984. *The Emergence of Dialectical Theory.* University of Chicago Press.

Watson, John. 1930. *Behaviorism.* New York: Norton.

Watson, John. 1913. "Psychology as the behaviorist views it." Pp. 158-177 in *Psychological Review,* 20.

Weaver, J., (ed.). 1987. *The World of Physics.* Vol. I-III. New York: Simon and Schuster.

Weibel, Ewald R. 1991. "Fractal Geometry: A Design Principle for Living Organisms." Pp. L361-L369 in *American Journal of Physiology.* V. 261:6

Weinberg, S. 1992. *Dreams of a Final Theory.* New York: Pantheon.

Weizsacher, C. F. von. 1980. *The Unity of Nature.* Trans. by F. Zucker. Farrar, Straus and Giroux.

West, Bruce J. 1996. "Fractal Statistics: Toward a Theory of Medicine." Pp. 263-295 in Fractal Horizons: *The Future Use of Fractals.* Ed. by C. Pickover. New York: St. Martin's Press.

Wheatley, D. and A. Unwin. 1972) *The Algorithm Writer's Guide.* London: Longman.

Wheelwright, P. 1959. *Heraclitus.* Princeton University Press

White, Elliott. 1992. *The End of the Empty Organism: Neurobiology and the Sciences of Human Action.* Westport, CN: Praeger.

Whitehead, Alfred N. 1929. *Process and Reality: An Essay in Cosmology.* New York: Macmillan.

Wigner, E. 1960. "The unreasonable effectiveness of mathematics in the natural sciences." in *Communications in Pure and Applied Mathematics.* V. 13. No. 1. New York: Wiley.

Wigoder, G. 1989. *The Encyclopedia of Judaism.* New York: MacMillan Publishing.

Wilhelm, H. 1960. *Change: Eight Lectures on the I Ching.* Trans. by C.F. Baynes. Pantheon.

Wilhelm, R. 1967. *The I Ching.* Trans. by C. Baynes. Princeton University Press.

Williams, R. J. P. and J. J. R. Frausto da Silva. 1996. *The Natural Selection of Chemical Elements.* Oxford: Clarendon Press.

Wilson, D. R. 1993. "Evolutionary Epidemiology: Darwinian Theory in the service of Medicine and Psychiatry." Pp. 205-218 in *Acta Biotheoretica,* 41.

Wilson, Edward O. 1998. *Consilience: The Unity of Knowledge.* New York: Alfred A. Knopf.

Wilson, Edward O. 1993. "Analyzing the Superorganism: The Legacy of Whitman and Wheeler." Pp. 243-255 in *The Biological Century.* Ed. by R. B. Barlow, Jr., J. E. Dowling, and G. Weissmann. Harvard University Press.

Wilson, Edward O. 1978. *On Human Nature.* New York: Bantam Books.

Wispe, Lauren. 1991. *The Psychology of Sympathy.* New York: Plenum Press.

Wolsky, M. and A. Wolsky. 1992. "Bergson's vitalism in the light of modern biology." in *The Crisis in Modernism.* Ed. by F. Burwick, and P. Douglas. Cambridge University Press.

Yankelovich, Daniel. 1981. *New Rules: Searching for Self-Fullfilment in a World Turned Upside Down.* New York: Random House.
Yogananda, Swami. 1994. *Autobiography of a Yogi.* Los Angeles, CA: Self-Realization Fellowship.
Zaehner, R. 1957. *Mysticism: Sacred and Profane.* Oxford University Press.
Zeki, S. 1993. *A Vision of the Brain.* London: Blackwell Scientific.

INDEX

MIX
Papier aus verantwortungsvollen Quellen
Paper from responsible sources
FSC® C105338

If you have any concerns about our products,
you can contact us on
ProductSafety@springernature.com

In case Publisher is established outside the EU,
the EU authorized representative is:
Springer Nature Customer Service Center GmbH
Europaplatz 3, 69115 Heidelberg, Germany

Printed by Libri Plureos GmbH
in Hamburg, Germany